GCSE Success

Workbook

Mathematics Higher

Fiona C. Mapp

Contents

Number

Algebra

Shape, Space and Measures

Handling Data

Homework diary

TOPIC	SCORE
Fractions	/29
Percentages 1	/32
Percentages 2	/33
Fractions, decimals and percentages	/30
Approximations and using a calculator	/31
Ratio	/24
Indices	/46
Standard index form	/31
Recurring decimals and surds	/46
Direct and inverse proportion	/31
Upper and lower bounds of measurement	/28
Algebra	/38
Equations	/38
Equations and inequalities	/35
Further algebra and equations	/37
Straight line graphs	/20
Curved graphs	/24
Harder work on graphs	/22
Interpreting graphs	/23
Bearings and scale drawings	/22
Transformations 1	/22
Transformations 2	/21
Similarity and congruency	/25
Loci and coordinates in 3D	/20
Angle properties of circles	/20
Pythagoras' theorem	/28
Trigonometry 1	/27
Trigonometry 2	/29
Further trigonometry	/37
Measures and measurement	/34
Area of 2D shapes	/32
Volume of 3D shapes	/29
Further length, area and volume	/26
Vectors	/26
Collecting data	/18
Scatter diagrams and correlation	/21
Averages 1	/29
Averages 2	/22
Cumulative frequency graphs	/25
Histograms	/14
Probability	/33

Planning and revising

- Mathematics should be revised **actively**. You should be doing **more than just reading**.

- Find out the dates of your first mathematics examination. Make an examination and revision timetable.

- After completing a topic in school, go through the topic again in the **GCSE Success Guide**. Copy out the **main points**, **results** and **formulae** into a notebook or use a **highlighter** to emphasise them.

- Try and write out the **key points** from **memory**. Check what you have written and see if there are any differences.

- Revise in short bursts of about **30 minutes**, followed by a **short break**.

- Learn **facts** from your exercise books, notebooks and the **Success Guide**. **Memorise** any formula you need to learn.

- Learn with a friend to make it easier and more fun!

- Do the **multiple choice** and **quiz-style** questions in this book and check your solutions to see how much you know.

- Once you feel **confident** that you know the topic, do the **GCSE**-style questions in this book. **Highlight** the key words in the question, **plan** your answer and then go back and **check** that you have answered the question.

- **Make a note** of any topics that you do not understand and **go back through** the notes again.

Different types of questions

- On the **GCSE Mathematics papers** you will have several types of questions:

  Calculate – In these questions you need to work out the answer. Remember that it is important to show full working out.

  Explain – These questions want you to explain, with a mathematical reason or calculation, what the answer is.

  Show – These questions usually require you to show, with mathematical justification, what the answer is.

  Write down or state – These questions require no explanation or working out.

  Prove – These questions want you to set out a concise logical argument, making the reasons clear.

  Deduce – These questions make use of an earlier answer to establish a result.

On the day

- **Follow the instructions** on the exam paper. Make sure that you understand what any **symbols** mean.

- Make sure that you **read the question** carefully so that you give the answer that an examiner wants.

- Always **show your working**; you may pick up some marks even if your final answer is wrong.

- Do **rough calculations** to check your answers and make sure that they are **reasonable**.

- When carrying out a calculation, **do not round the answer until the end**, otherwise your final answer will not be as accurate as is needed.

- Lay out your working **carefully** and **concisely**. Write down the calculations that you are going to make. You usually get marks for showing a **correct method**.

- Make your drawings and graphs **neat** and **accurate**.

- Know what is on the **formula sheet** and make sure that you **learn** those formulae that are not on it.

- If you cannot do a question, **leave it out** and **go back** to it at the end.

- Keep an eye on the time. Allow enough time to check through your answers.

- If you finish early, check through everything very carefully and try and fill in any gaps.

- Try and write something even if you are not sure about it. Leaving an empty space will score you no marks.

In this book, questions which may be answered with a calculator are marked with Ⓒ. All the other questions are intended to be answered without the use of a calculator.

Good luck!

Fractions

A   Choose just one answer, a, b, c or d.

1 Which one of these fractions is equivalent to $\frac{5}{9}$?   **(1 mark)**

a) $\frac{16}{27}$ b) $\frac{9}{18}$

c) $\frac{25}{45}$ d) $\frac{21}{36}$

2 In a class of 24 students, $\frac{3}{8}$ wear glasses. How many students wear glasses?   **(1 mark)**

a) 9
b) 6
c) 3
d) 12

3 Work out the answer to $1\frac{5}{9} - \frac{1}{3}$ **(1 mark)**

a) $\frac{1}{3}$ b) $1\frac{2}{9}$

c) $1\frac{4}{6}$ d) $\frac{4}{12}$

4 Work out the answer to $\frac{2}{11} \times \frac{7}{9}$ **(1 mark)**

a) $\frac{14}{11}$ b) $\frac{14}{9}$

c) $\frac{14}{99}$ d) $\frac{2}{99}$

5 Work out the answer to $\frac{3}{10} \div \frac{2}{3}$ **(1 mark)**

a) $\frac{9}{20}$ b) $\frac{6}{50}$

c) $\frac{6}{15}$ d) $\frac{4}{3}$

Score   / 5

B   Answer all parts of the questions.

1 Work out the answers to the following.   **(8 marks)**

a) $\frac{2}{9} + \frac{1}{3}$ ....   b) $\frac{7}{11} - \frac{1}{4}$ ....   c) $\frac{4}{7} \times \frac{3}{8}$ ....

d) $\frac{9}{12} \div \frac{1}{4}$ ....   e) $2\frac{5}{7} - 1\frac{1}{21}$ ....   f) $\frac{4}{9} + \frac{3}{27}$ ....

g) $\frac{7}{12} \times 1\frac{1}{2}$ ....   h) $1\frac{4}{7} \div \frac{7}{12}$ ....

2 Arrange these fractions in order of size, smallest first.   **(2 marks)**

a) $\frac{2}{3}$ $\frac{4}{5}$ $\frac{1}{7}$ $\frac{3}{4}$ $\frac{1}{2}$ $\frac{3}{10}$

....

b) $\frac{5}{8}$ $\frac{1}{3}$ $\frac{2}{7}$ $\frac{1}{9}$ $\frac{3}{4}$ $\frac{2}{5}$ **(2 marks)**

....

3 Decide whether these statements are true or false.

a) $\frac{4}{5}$ of 20 is bigger than $\frac{6}{7}$ of 14.   ....   **(1 mark)**

b) $\frac{2}{9}$ of 27 is smaller than $\frac{1}{3}$ of 15.   ....   **(1 mark)**

4 In a class of 32 pupils, $\frac{1}{8}$ are left-handed. How many students are not left-handed?   **(1 mark)**

....

Score   / 15

C These are GCSE-style questions. Answer all parts of the questions. Show your workings (on separate paper if necessary) and include the correct units in your answers.

1 Gill says 'I've got three-fifths of a bottle of orange juice.'

Jonathan says 'I've got two-thirds of a bottle of orange juice and my bottle of orange juice is the same size as yours.'

Who has the most orange juice, Gill or Jonathan? Explain your answer. (2 marks)

....

....

2 Work out these.

a) $\frac{2}{3} + \frac{4}{5}$ .... (1 mark)

b) $3\frac{9}{11} - 2\frac{1}{3}$ .... (1 mark)

c) $\frac{2}{7} \times \frac{4}{9}$ .... (1 mark)

d) $\frac{3}{10} \div \frac{2}{5}$ .... (1 mark)

3 Charlotte's take home pay is £930. She gives her mother $\frac{1}{3}$ of this and spends $\frac{1}{5}$ of the £930 on going out.

What fraction of the £930 is left? Give your answer as a fraction in its lowest terms. (3 marks)

....

....

Score / 9

How well did you do? ✗ 1–11 Try again 12–18 Getting there 19–24 Good work 25–29 Excellent! ✓

For more information on this topic, see pages 8–9 of your Success Guide.

Percentages 1

A

Choose just one answer, a, b, c or d.

1 Work out 10% of £850. (1 mark)

a) £8.50
b) £0.85
c) £85
d) £42.50

2 Work out 17.5% of £60. (1 mark)

a) £9
b) £15
c) £10.50
d) £12.50

3 In a survey, 17 people out of 25 said they preferred type A cola. What percentage of people preferred type A cola? (1 mark)

a) 68%   b) 60%
c) 72%   d) 75%

4 A CD player costs £60 in a sale after a reduction of 20%. What was the original price of the CD player? (c) (1 mark)

a) £48
b) £70
c) £72
d) £75

5 A new car was bought for £15 000. Two years later it was sold for £12 000. What is the percentage loss? (c) (1 mark)

a) 25%
b) 20%
c) 80%
d) 70%

Score   / 5

B

Answer all parts of the questions.

1 Work out the answer to the following. (4 marks)

a) 20% of £60 ....   b) 30% of £150 ....
c) 5% of £80 ....   d) 12.5% of 40 g ....

2 Colin earns £25 500 a year. This year he has a 3% pay rise. How much does Colin now earn? (c)

£ .... (2 marks)

3 A coat costs £140. In a sale it is reduced to £85. What is the percentage reduction? (c) (2 marks)

.... %

4 Lucinda scored 58 out of 75 in a test. What percentage did she get? (c) (2 marks)

.... %

5 The cost for a ticket for a pop concert has risen by 15% to £23. What was the original price of the ticket? (c) (2 marks)

£ ....

6 The price of a CD player has been reduced by 20% in a sale. It now costs £180. What was the original price? (c) (2 marks)

£ ....

7 12 out of 30 people wear glasses. What percentage wear glasses? (c) (2 marks)

.... %

Score   / 16

 Indicates that a calculator may be used

C

These are GCSE-style questions. Answer all parts of the questions. Show your workings (on separate paper if necessary) and include the correct units in your answers.

1 The price of a television set is £175 plus VAT. VAT is charged at a rate of 17.5%. Ⓒ

a) Work out the amount of VAT charged.

(2 marks)

....

b) In a sale, normal prices are reduced by 15%. The normal price of a washing machine is £399.

Work out the sale price of the washing machine.

(3 marks)

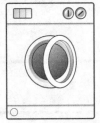

....

2 A car is bought for £17 900. Two years later it is sold for £14 320. Work out the percentage loss. Ⓒ

(3 marks)

....

3 In a sale, all normal prices are reduced by 18%. In the sale, Suki pays £57.40 for a jacket.

Calculate the normal price of the jacket. Ⓒ

(3 marks)

....

Score   / 11

How well did you do? ✗ 1–7 Try again   8–17 Getting there   18–26 Good work   27–32 Excellent! ✓

For more information on this topic, see pages 12–13 of your Success Guide.

9

Percentages 2

A Choose just one answer, a, b, c or d.

1 £2000 is invested in a savings account. Compound interest is paid at 2.1%. How much interest is paid after 2 years? (c)   (1 mark)

a) £4
b) £5.20
c) £2.44
d) £84.88

2 A bike was bought for £120. Each year it depreciated by 10%. What was the bike worth 2 years later? (c)   (1 mark)

a) £97.20
b) £98
c) £216
d) £110

3 Roberto has £5000 in his savings account. Simple interest is paid at 3%. How much does he have in his savings account at the end of the year? (c)   (1 mark)

a) £4850
b) £5010
c) £5150
d) £5140.50

4 Lily earns £23 500. National Insurance (NI) is deducted at 9%. How much NI must she pay?

a) £2250   (c) (1 mark)
b) £2115
c) £2200
d) £21 385

Score   / 4

B Answer all parts of the questions.

1 A meal costs £143. VAT at 17.5% is added to the price of the meal. What is the final price of the meal? (c)   (2 marks)

£ ....

2 VAT of 5% is added to a gas bill of £72. Find the total amount to be paid. (c)   (2 marks)

£ ....

3 A motorbike is bought for £9000. Each year it depreciates in value by 12%. Work out the value of the motorbike after 2 years. (c)   (2 marks)

£ ....

4 Scarlett has £6200 in her savings account. If compound interest is paid at 2.7% p.a., how much interest will she earn after 3 years? (c)   (2 marks)

£ ....

5 A house was bought for £112 000. After the first year the price had increased by 8%, during the second year it increased in price by a further 12%. What is the house now worth? (c)   (2 marks)

£ ....

6 Petrol cost 74.9 pence per litre. The price increased by 2%. Six months later it increased again, by 5%. How much does a litre of petrol now cost? (c)   (2 marks)

.... pence

Score   / 12

(c) *Indicates that a calculator may be used*

C

These are GCSE-style questions. Answer all parts of the questions. Show your workings (on separate paper if necessary) and include the correct units in your answers.

1 a) Work out 40% of £2500. (2 marks)

....

b) Find the simple interest on £2000 invested for 2 years at 4% per year. (3 marks)

....

2 £7000 is invested for 3 years at 6% compound interest. Work out the total interest earned over the three years. Ⓒ (3 marks)

....

3 Nigel opened an account with £450 at his local bank. After one year, the bank paid him interest. He then had £465.75 in his account.

a) Work out, as a percentage, his local bank's interest rate. Ⓒ (3 marks)

....

b) Sarah opened an account at the same bank as Nigel. She invested £700 for 2 years at 4% compound interest. How much money did she have in her account after 2 years? (3 marks)

....

4 A vintage bottle of champagne was valued at £42 000 on 1 January 2005.

The value of the champagne is predicted to increase at a rate of R% per annum.
The predicted value, £V, of the champagne after n years is given by the formula Ⓒ

$$V = 42\,000 \times (1.045)^n$$

a) Write down the value of R. (1 mark)

....

b) Use your calculator to find the predicted value of the champagne after 8 years. (2 marks)

....

Score   / 17

How well did you do? ✗ 1–7 Try again   8–14 Getting there   15–24 Good work   25–33 Excellent! ✓

For more information on this topic, see pages 12–15 of your Success Guide.

11

Fractions, decimals, percentages

A Choose just one answer, a, b, c or d.

1 What is $\frac{3}{5}$ as a percentage? (1 mark)

a) 30%   b) 25%

c) 60%   d) 75%

2 What is $\frac{2}{3}$ written as a decimal? (1 mark)

a) 0.77   b) $0.\dot{6}$

c) 0.665   d) 0.6

3 What is the smallest value in this list of numbers?   **29%   0.4 $\frac{3}{4}$ $\frac{1}{8}$** (1 mark)

a) 29%   b) 0.4   c) $\frac{3}{4}$ d) $\frac{1}{8}$

4 What is the largest value in this list of numbers? $\frac{4}{5}$ 80% $\frac{2}{3}$ 0.9 (1 mark)

a) $\frac{4}{5}$ b) 80%

c) $\frac{2}{3}$ d) 0.9

5 Change $\frac{5}{8}$ into a decimal. (1 mark)

a) 0.625   b) 0.425

c) 0.125   d) 0.725

Score   / 5

B Answer all parts of the questions.

1 The table shows equivalent fractions, decimals and percentages. Fill in the gaps. (2 marks)

Fraction	Decimal	Percentage
$\frac{2}{5}$		
		5%
	$0.\dot{3}$	
	0.04	
		25%
$\frac{1}{8}$		

2 Put these cards in order of size, smallest first. (6 marks)

| 0.37 | 30% | $\frac{3}{8}$ | $\frac{1}{3}$ | 92% | $\frac{1}{2}$ | 0.62 |

( ) ( ) ( ) ( ) ( ) ( ) ( )

3 Decide whether these calculations give the same answer for this question: **Increase £40 by 20%**

Jack says: Multiply 40 by 1.2

Hannah says: Work out 10%, double it and then add 40

Explain your reasoning.

.... (2 marks)

....

Score   / 10

C

These are GCSE-style questions. Answer all parts of the questions. Show your workings (on separate paper if necessary) and include the correct units in your answers.

1 Write this list of seven numbers in order of size. Start with the smallest number. (3 marks)

25% $\frac{1}{3}$ 0.27 $\frac{2}{5}$ 0.571   72% $\frac{1}{8}$

....

2 Philippa is buying a new television. She sees three different advertisements for the same television set. **ⓒ**

Ed's Electricals

TV normal price

£250

Sale 10% off

Sheila's Bargains

TV **£185** plus

VAT at $17\frac{1}{2}$%

GITA's TV SHOP

Normal price

£290

Sale: $\frac{1}{5}$ off normal price

a) Find the maximum and minimum prices that Bronwen could pay for a television set. (7 marks)

Maximum price = ...

Minimum price = ...

b) The price of the television in a fourth shop is £235. This includes VAT at 17.5%. Work out the cost of the television before VAT was added. (3 marks)

....

3 A sundial is being sold in two different garden centres. The cost of the sundial is £89.99 in both garden centres. Both garden centres have a promotion.

Gardens are Us   Sundial 22% off

Rosebushes   Sundial $\frac{1}{4}$ off

In which garden centre is the sundial cheaper? Explain your reasoning. (2 marks)

....

....

Score   / 15

How well did you do?   ✗ **1–7** Try again   **8–13** Getting there   **14–22** Good work   **23–30** Excellent! ✓

For more information on this topic, see page 16 & 12–13 of your Success Guide.

13

Approximations & using a calculator

A Choose just one answer, a, b, c or d.

1 Estimate the answer to the calculation
27 × 41 (1 mark)

a) 1107   b) 1200
c) 820   d) 1300

2 A carton of orange juice costs 79p. Estimate the cost of 402 cartons of orange juice. (1 mark)

a) £350
b) £250
c) £400
d) £320

3 A school trip is organised. 407 pupils are going on the trip. Each coach seats 50 pupils. Approximately how many coaches are needed?

a) 12   b) 5   (1 mark)
c) 8   d) 10

4 Estimate the answer to the calculation $\frac{(4.2)^2}{107}$

a) 16   b) 1.6   (1 mark)
c) 0.16   d) 160

5 Round 5379 to 3 significant figures.

a) 538   b) 5370   (1 mark)
c) 537   d) 5380

Score   / 5

B Answer all parts of the questions.

1 Decide whether each statement is true or false. (4 marks)

a) 2.742 rounded to 3 significant figures is 2.74   ....

b) 2793 rounded to 2 significant figures is 27   ....

c) 32 046 rounded to 1 significant figure is 40 000   ....

d) 14.637 rounded to 3 significant figures is 14.6   ....

2 Round each of the numbers in the calculations to 1 significant figure then work out an approximate answer.

a) $\frac{(32.9)^2}{9.1}$ .... (1 mark)

b) $\frac{(906 \div 31.4)^2}{7.1 + 2.9}$ .... (1 mark)

3 Work these out on your calculator. Give your answers to 3 s.f. 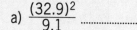 (4 marks)

a) $\frac{4.2(3.6 + 5.1)}{2 - 1.9}$ ....

b) $6 \times \sqrt{\frac{12.1}{4.2}}$ ....

c) $\frac{12^5}{4.3 \times 9.15}$ ....

d) $\frac{4\cos 30° + 2\sin 60°}{4^3}$ ....

Score   / 10

C *Indicates that a calculator may be used*

Number

C These are GCSE-style questions. Answer all parts of the questions. Show your workings (on separate paper if necessary) and include the correct units in your answers.

1 a) Write down two numbers you could use to get an approximate answer to this question. (1 mark)

  31×79 .... and ....

  b) Work out your approximate answer. (1 mark)

  ....

  ....

  c) Work out the difference between your approximate answer and the exact answer. (2 marks)

  ....

  ....

  ....

2 Use your calculator to work out the value of this. (C)

  $$\frac{\sqrt{(4.9^2 + 6.3)}}{2.1 \times 0.37}$$

  Give your answer correct to 3 significant figures. (3 marks)

  ....

3 a) Use your calculator to work out the value of this. (C)

  $$\frac{27.1 \times 6.2}{38.2 - 9.9}$$

  Write down all the figures on your calculator display. (2 marks)

  ....

  b) Round each of the numbers in the above calculation to 1 significant figure and obtain an approximate answer. (3 marks)

  ....

4 Circle the best estimate for each of these calculations. (4 marks)

	A	B	C
a) 52×204	700	10 000	1000
b) $904 \div 31$	320	30	300
c) 1.279×4.9	4	6	8
d) $2795 \div 19.1$	150	195	102

Score   / 16

How well did you do? ✗ 1–11 Try again   12–19 Getting there   20–26 Good work   27–31 Excellent! ✓

For more information on this topic, see pages 17–19 of your Success Guide.

Ratio

A Choose just one answer, a, b, c or d.

1 What is the ratio 6 : 18 written in its simplest form? *(1 mark)*

a) 3 : 1   b) 3 : 9
c) 1 : 3   d) 9 : 3

2 Write the ratio 200 : 500 in the form 1 : n.

a) 1 : 50   b) 1 : 5   *(1 mark)*
c) 1 : 25   d) 1 : 2.5

3 If £140 is divided in the ratio 3 : 4, what is the size of the larger share? *(1 mark)*

a) £45   b) £60
c) £80   d) £90

4 A recipe for 4 people needs 800 g of flour. How much flour is needed for 6 people?

a) 12 g   b) 120 g   *(1 mark)*
c) 12 kg   d) 1200 g

5 If 9 oranges cost £1.08, how much would 14 similar oranges cost? *(1 mark)*

a) £1.50   b) £1.68
c) £1.20   d) £1.84

Score   / 5

B Answer all parts of the questions.

1 Write down each of the following ratios in the form 1 : n.

a) 10 : 15 ....   *(1 mark)*

b) 6 : 10 ....   *(1 mark)*

c) 9 : 27 ....   *(1 mark)*

2 Seven bottles of lemonade have a total capacity of 1680 ml. Work out the total capacity for five similar bottles.

.... ml   *(1 mark)*

3 a) Increase £4.10 in the ratio 2 : 5 ....   *(1 mark)*

b) Decrease 120 g in the ratio 5 : 2 ....   *(1 mark)*

4 Mrs London inherited £55 000. She divided the money between her children in the ratio 3 : 3 : 5. How much did the child with the largest share receive? *(2 marks)*

£ ....

5 It takes 6 people 3 days to dig and lay a cable. How long would it take 4 people? *(2 marks)*

.... days

Score   / 10

C These are GCSE-style questions. Answer all parts of the questions. Show your workings (on separate paper if necessary) and include the correct units in your answers.

1 Vicky and Tracy share £14 400 in the ratio 4 : 5. Work out how much each of them receives.

Vicky: £ .... Tracy: £ .... (3 marks)

2 James uses these ingredients to make 12 buns.

> 50 g butter
> 40 g sugar
> 2 eggs
> 45 g flour
> 15 ml milk

James wants to make 18 similar buns. Write down how much of each ingredient he needs for 18 buns. (3 marks)

butter .... g   sugar .... g

eggs ....   flour .... g

milk .... ml

3 It takes 3 builders 16 days to build a wall. All the builders work at the same rate. How long would it take 8 builders to build a wall the same size? (3 marks)

.... days

Score   / 9

How well did you do? ✗ 1–6 Try again   7–11 Getting there   12–17 Good work   18–24 Excellent! ✓

For more information on this topic, see pages 20–21 of your Success Guide.

17

RATIO   Number

Indices

A
Choose just one answer, a, b, c or d.

1 In index form, what is the value of $8^3 \times 8^{11}$?

   a) 8^{14} b) 8^{33} (1 mark)
   c) 64^{14} d) 64^{33}

2 In index form, what is the value of $(4^2)^3$?

   a) 12^2 b) 4^5 (1 mark)
   c) 4^6 d) 16^6

3 What is the value of 5^0?   (1 mark)

   a) 5   b) 0
   c) 25   d) 1

4 What is the value of 5^{-2}?   (1 mark)

   a) $\frac{1}{25}$ b) -5
   c) 25   d) -25

5 What is the value of $7^{-12} \div 7^2$ written in index form?   (1 mark)

   a) 7^{10} b) 7^{-14}
   c) 7^{14} d) 7^{-10}

Score   / 5

B
Answer all parts of the questions.

1 Decide whether each of these expressions is true or false.

                                                      True   False

   a) $a^4 \times a^5 = a^{20}$ ☐   ☐   (1 mark)

   b) $2a^4 \times 3a^2 = 5a^8$ ☐   ☐   (1 mark)

   c) $10a^6 \div 2a^4 = 5a^2$ ☐   ☐   (1 mark)

   d) $20a^4b^2 \div 10a^5b = 2a^{-1}b$ ☐   ☐   (1 mark)

   e) $(2a^3)^3 = 6a^9$ ☐   ☐   (1 mark)

   f) $4^0 = 1$ ☐   ☐   (1 mark)

2 Simplify the following expressions.   (4 marks)

   a) $(5a)^0 =$ ....   b) $(2a^2)^4 =$ ....

   c) $12a^4 \div 16a^7 =$ ....   d) $(3a^2b^3)^3 =$ ....

3 Write these using negative indices.   (3 marks)

   a) $\frac{4}{x^2} =$ ....   b) $\frac{a^2}{b^3} =$ ....   c) $\frac{3}{y^5} =$ ....

4 Evaluate these expressions.   (4 marks)

   a) $25^{-\frac{1}{2}}$ b) $49^{\frac{3}{2}}$ c) $\left(\frac{4}{5}\right)^{-2}$ d) $81^{\frac{-3}{4}}$

   ....

Score   / 17

C

These are GCSE-style questions. Answer all parts of the questions. Show your workings (on separate paper if necessary) and include the correct units in your answers.

1 Simplify these.

a) $p^3 \times p^4$ .... (1 mark)

b) $\dfrac{n^3}{n^7}$ .... (1 mark)

c) $\dfrac{a^3 \times a^4}{a}$ .... (1 mark)

d) $\dfrac{12a^2b}{3a}$ .... (1 mark)

2 Work out these. (5 marks)

a) 3^0 .... b) 9^{-2} .... c) $3^4 \times 2^3$ ....

d) $64^{\frac{2}{3}}$ .... e) $125^{-\frac{1}{3}}$ ....

3 a) Evaluate the following.

i) 8^0 .... (1 mark)

ii) 4^{-2} .... (1 mark)

iii) $\left(\dfrac{4}{9}\right)^{-\frac{1}{2}}$ .... (1 mark)

b) Write this as a single power of 5.

$\dfrac{5^7 \times 5^3}{(5^2)^3}$ .... (2 marks)

4 Evaluate the following, giving your answers as fractions.

a) 5^{-3} .... (1 mark)

b) $\left(\dfrac{2}{3}\right)^{-2}$ .... (1 mark)

c) $(8)^{-\frac{2}{3}}$ .... (1 mark)

5 Simplify these, leaving your answer in the form 2^n.

a) $4^{-\frac{1}{2}}$ .... (1 mark)

b) $\dfrac{2^7 \times 2^9}{2^{-4}}$ .... (1 mark)

c) $(\sqrt{2})^5$ .... (1 mark)

6 Simplify these expressions. (4 marks)

a) $(5x)^{-3}$ b) $(2x^2y^3)^{-4}$ c) $(3y)^{-3}$ d) $(2xy^3)^5$

....

Score   / 24

How well did you do? ✗ 1–12 Try again   13–23 Getting there   24–36 Good work   37–46 Excellent! ✓

For more information on this topic, see pages 22–23 of your Success Guide.

Standard index form

A Choose just one answer, a, b, c or d.

1 What is this number written in standard form?
42 710 *(1 mark)*

a) 42.71×10^3 b) 4.271×10^4
c) 4271.0×10 d) 427.1×10^2

2 What is 6.4×10^{-3} written as an ordinary number? *(1 mark)*

a) 6400   b) 0.0064
c) 64   d) 0.064

3 What is 2.7×10^4 written as an ordinary number? *(1 mark)*

a) 27 000   b) 0.27
c) 270   d) 0.00027

4 What would $(4 \times 10^9) \times (2 \times 10^6)$ worked out and written in standard form be?

a) 8×10^{54} b) 8×10^{15} *(1 mark)*
c) 8×10^3 d) 6×10^{15}

5 What would $(3 \times 10^4)^2$ worked out and written in standard form be? *(1 mark)*

a) 9×10^6 b) 9×10^8
c) 9×10^9 d) 3×10^8

Score   / 5

B Answer all parts of the questions.

1 Decide whether each of the statements is true or false.

a) 4710 is 4.71×10^3 written in standard form. .... *(1 mark)*

b) 249 000 is 24.9×10^4 written in standard form. .... *(1 mark)*

c) 0.047 is 47×10^{-3} written in standard form. .... *(1 mark)*

d) 0.000 009 6 is 9.6×10^{-7} written in standard form. .... *(1 mark)*

2 Carry out the following calculations. Give your answers in standard form.

a) $(4 \times 10^6) \times (2 \times 10^9)$ .... *(1 mark)*

b) $(7 \times 10^{-3}) \times (2 \times 10^6)$ .... *(1 mark)*

c) $(9 \times 10^{12}) \div (3 \times 10^{-4})$ .... *(1 mark)*

d) $(2.4 \times 10^{10}) \div (3 \times 10^6)$ .... *(1 mark)*

3 Work out these calculations. Give your answer in standard form. Ⓒ

a) $(2.1 \times 10^7) \times (3.9 \times 10^{-4})$ .... *(1 mark)*

b) $(6.3 \times 10^{-4}) \times (1.2 \times 10^7)$ .... *(1 mark)*

c) $(1.2 \times 10^{-7}) \div (2 \times 10^{-4})$ .... *(1 mark)*

d) $(8.9 \times 10^6) \div (4 \times 10^{-2})$ .... *(1 mark)*

4 The mass of an atom is 2×10^{-23} grams. *(3 marks)*

What is the total mass of 9×10^{15} of these atoms? .... Ⓒ

Score   / 15

Ⓒ *Indicates that a calculator may be used*

C These are GCSE-style questions. Answer all parts of the questions. Show your workings (on separate paper if necessary) and include the correct units in your answers.

1 a) i) Write the number 2.07×10^5 as an ordinary number. (2 marks)

....

ii) Write the number 0.000 046 in standard form.

....

b) Multiply 7×10^4 by 5×10^7

Give your answer in standard form. (2 marks)

....

2 Calculate the value of $\dfrac{4.68 \times 10^9 + 3.14 \times 10^7}{2.14 \times 10^{-3}}$

Give your answer in standard form, correct to 2 significant figures. Ⓒ (3 marks)

....

3 3.8×10^8 seeds weigh 1 kilogram.

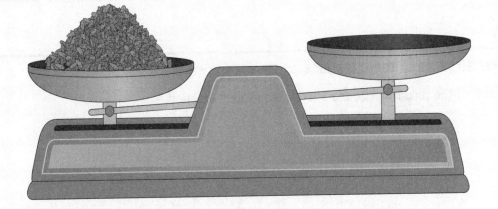

Each seed weighs the same. Calculate the weight in grams of one seed.
Give your answer in standard form, correct to 2 significant figures. Ⓒ (2 marks)

.... g

4 If $a = 3.2 \times 10^4$ and $b = 2 \times 10^{-3}$, calculate the answer to $\dfrac{b^2}{a+b}$

giving your answer in standard form, correct to 3 significant figures. Ⓒ (2 marks)

....

Score   / 11

How well did you do? ✗ 1–9 **Try again**   10–19 **Getting there**   20–26 **Good work**   27–31 **Excellent!** ✓

For more information on this topic, see pages 24–25 of your Success Guide.

21

Recurring decimals & surds

A Choose just one answer, a, b, c or d.

1 Which fraction is the same as $0.\dot{5}$? (1 mark)

 a) $\frac{1}{2}$ b) $\frac{5}{10}$

 c) $\frac{5}{9}$ d) $\frac{5}{8}$

2 Which fraction is equivalent to $0.\dot{6}\dot{3}...$?

 a) $\frac{63}{100}$ b) $\frac{6}{99}$ (1 mark)

 c) $\frac{636}{999}$ d) $\frac{7}{11}$

3 Which fraction is equivalent to $0.2\dot{1}...$?

 a) $\frac{19}{90}$ b) $\frac{21}{99}$ (1 mark)

 c) $\frac{211}{999}$ d) $\frac{2}{9}$

4 Which expression is equivalent to $\sqrt{12}$?

 a) $2\sqrt{6}$ b) $2\sqrt{3}$ (1 mark)

 c) $6\sqrt{2}$ d) $3\sqrt{2}$

5 Which expression is equivalent to $\frac{1}{\sqrt{3}}$? (1 mark)

 a) $\frac{\sqrt{3}}{3}$ b) $\frac{\sqrt{3}}{9}$

 c) $\frac{9}{\sqrt{3}}$ d) $\frac{3}{\sqrt{3}}$

Score / 5

B Answer all parts of the questions.

1 Match each of the recurring decimals to the equivalent fraction. (5 marks)

$0.\dot{3}$ $\quad\quad$ $\frac{7}{9}$

$0.\dot{7}$ $\quad\quad$ $\frac{244}{333}$

$0.2\dot{4}$ $\quad\quad$ $\frac{13}{30}$

$0.\dot{7}3\dot{2}$ $\quad\quad$ $\frac{1}{3}$

$0.4\dot{3}$ $\quad\quad$ $\frac{8}{33}$

2 Find the fraction which is equivalent to $0.1\dot{2}\dot{5}$. Express the fraction in its simplest form. (2 marks)

....

....

3 Express each of the following in the form $a\sqrt{b}$, where a and b are integers and b is as small as possible.

 a) $\sqrt{24}$ .... (1 mark)

 b) $\sqrt{75}$ .... (1 mark)

 c) $\sqrt{48}+\sqrt{12}$ .... (2 marks)

 d) $\sqrt{80}+\sqrt{20}$ .... (2 marks)

Score / 13

C These are GCSE-style questions. Answer all parts of the questions. Show your workings (on separate paper if necessary) and include the correct units in your answers.

1 a) Change the decimal $0.5\dot{4}$ into a fraction in its lowest terms. (2 marks)

....

b) Write the recurring decimal $0.0\dot{2}\dot{6}$ as a fraction. (2 marks)

....

2 a) Find the value of $\sqrt{3} \times \sqrt{27}$ .... (1 mark)

b) $\sqrt{3} + \sqrt{27} = a\sqrt{3}$, where a is an integer.

Find the value of a. (1 mark)

....

c) Find the value of $\dfrac{\sqrt{3}+\sqrt{12}}{\sqrt{75}}$ (3 marks)

....

3 Work out $\dfrac{(5+\sqrt{5})(2-2\sqrt{5})}{\sqrt{45}}$

Give your answer in its simplest form. (3 marks)

....

4 Write down the recurring decimal $0.12\dot{3}$ in the form $\frac{a}{b}$ where a and b are integers. (2 marks)

....

5 Simplify $(4 - \sqrt{3})^2$ (2 marks)

....

6 Express $\dfrac{\sqrt{125}+\sqrt{50}}{\sqrt{5}}$ in the form $a + \sqrt{b}$ (4 marks)

7 Work out $\dfrac{(2-\sqrt{2})(4+3\sqrt{2})}{2}$ (3 marks)

Give your answer in the form $a + b\sqrt{c}$

....

8 Write down the recurring decimal $0.7\dot{5}$ as a fraction in its simplest form. (2 marks)

....

9 Prove that the recurring decimal $0.4\dot{5}$ is $\frac{5}{11}$. (3 marks)

....

How well did you do? ✗ 1–13 Try again   14–25 Getting there   26–38 Good work   39–46 Excellent! ✓

For more information on this topic, see pages **27** of your Success Guide.

23

Direct and inverse proportion

A

Choose just one answer, a, b, c or d.

1 If a is directly proportional to b and $a = 10$ when $b = 5$, what is the formula that connects a and b? (1 mark)

a) $a = 5b$
b) $a = 10b$
c) $a = \frac{1}{2}b$
d) $a = 2b$

2 If y is directly proportional to x and $y = 12$ when $x = 4$, what is the formula that connects x and y? (1 mark)

a) $y = 3x$
b) $x = 3y$
c) $y = 12x$
d) $y = \frac{1}{3x}$

3 If d is inversely proportional to c, so that $d = \frac{k}{c}$ and $d = 6$ when $c = 3$, what is the value of k? (1 mark)

a) 12
b) 18
c) 2
d) 9

4 If v is inversely proportional to w^2 and $v = 3$ when $w = 2$, what is the formula that connects v and w^2? (1 mark)

a) $v = \frac{18}{w^2}$
b) $v = \frac{2}{w^2}$
c) $v = \frac{12}{w^2}$
d) $v = \frac{3}{2w^2}$

Score / 4

B

Answer all parts of the questions.

1 Given that $a \propto b$, calculate the values missing from this table. (2 marks)

a			12	30
b		2		6

2 The variables x and y are related so that y is directly proportional to the square of x. Complete this table for values of x and y. (3 marks)

x	2	4		
y	12		27	75

3 z is inversely proportional to the square of v.

a) Express z in terms of v and a constant of proportionality k. (2 marks)

....

b) If $z = 10$ when $v = 5$, calculate

i) the value of z when $v = 2$ .... (2 marks)

ii) the value of v when $z = 1000$ .... (2 marks)

Score / 11

C *Indicates that a calculator may be used*

C These are GCSE-style questions. Answer all parts of the questions. Show your workings (on separate paper if necessary) and include the correct units in your answers.

1 The extension E of a spring is directly proportional to the force F pulling the spring.
The extension is 6 cm when a force of 15 N is pulling it.
Calculate the extension when the force is 80 N. (4 marks)

....

....

2 The volume (V) of a toy is proportional to the cube of its height (h).
When the toy's volume is 60 cm³ the height is 2 cm.
Find the volume of a similar toy whose height is 5 cm. **C** (4 marks)

....

....

3 I is inversely proportional to the square of d.
When d equals 2, I equals 50.

a) Calculate the value of I when d is 3.5. **C** (3 marks)

....

....

b) Calculate the value of d when I equals 12.5. (3 marks)

....

....

4 c is inversely proportional to b and $b = 10$ when $c = 4$. (2 marks)

Daisy says that the formula connecting c and b is given by $c = \frac{40}{b}$.

Decide whether Daisy is correct, giving a reason for your answer.

....

....

Score   / 16

How well did you do?   ✗ 1–6 **Try again**   7–13 **Getting there**   14–21 **Good work**   22–31 **Excellent!** ✓

For more information on this topic, see page 26 of your Success Guide.

25

Upper & lower bounds of measurement

A Choose just one answer, a, b, c or d.

1 The length of an object is **5.6 cm**, correct to the nearest millimetre. What is the lower bound of the length of the object? **(1 mark)**

a) 5.56 cm   b) 5.55 cm
c) 5.64 cm   d) 5.65 cm

2 The weight of an object is **2.23 grams**, correct to two decimal places. What is the upper bound of the weight of the object?

a) 2.225 g   b) 2.234 g   **(1 mark)**
c) 2.235 g   d) 2.32 g

3 A hall can hold **40 people** to the nearest ten. What is the upper boundary for the number of people in the hall? **(1 mark)**

a) 45   b) 35
c) 44   d) 36

4 A square has a length of **3 cm** to the nearest centimetre. What is the lower bound for the perimeter of the square? **(1 mark)**

a) 12 cm   b) 10 cm
c) 10.4 cm   d) 14 cm

5 Using the information given in the previous question, what is the upper bound for the area of the square? **(1 mark)**

a) 6.25 cm^2 b) 9 cm^2
c) 12.5 cm^2 d) 12.25 cm^2

Score   / 5

B Answer all parts of the questions.

1 Complete the inequalities in the questions below, which show the upper and lower bounds of some measurements.

a) .... $\leq 5.24 < 5.245$ (1 mark)

b) $3.5 \leq 4 <$ ....   (1 mark)

c) $0.3235 \leq 0.324 <$ ....   (1 mark)

d) $8.435 \leq$ .... < 8.445 (1 mark)

2 $a = \dfrac{(3.4)^2 \times 12.68}{2.4}$

3.4 and 2.4 are correct to 1 decimal place.
12.68 is correct to 2 decimal places.
Which of the following calculations gives the lower bound for a and the upper bound for a?
(Write down the letters.)   (2 marks)

a) $\dfrac{(3.45)^2 \times 12.685}{2.35}$ b) $\dfrac{(3.35)^2 \times 12.675}{2.35}$ c) $\dfrac{(3.45)^2 \times 12.685}{2.45}$

d) $\dfrac{(3.45)^2 \times 12.675}{2.35}$ e) $\dfrac{(3.35)^2 \times 12.675}{2.45}$

Lower bound ....   Upper bound ....   Score   / 6

C *Indicates that a calculator may be used*

C These are GCSE-style questions. Answer all parts of the questions. Show your workings (on separate paper if necessary) and include the correct units in your answers.

1 $p = 3.1$ cm and $q = 4.7$ cm, correct to one decimal place.

a) Calculate the upper bound for the value of $p + q$ Ⓒ (2 marks)

....

....

b) Calculate the lower bound for the value of $\frac{p}{q}$ (3 marks)

Give your answer correct to 3 significant figures.

....

....

....

2 The volume of a cube is given as 62.7 cm³, correct to 1 decimal place. Find the upper and lower bounds for the length of an edge of this cube. Ⓒ (4 marks)

Lower bound = ...

Upper bound = ...

3 The mass of an object is measured as 120 g, and its volume as 630 cm³. Both of these measurements are correct to 2 significant figures. Find the range of possible values for the density of the object. Ⓒ (4 marks)

....

....

4 To the nearest centimetre, $a = 3$ cm, $b = 5$ cm.

Calculate the lower bound for ab. .... cm² Ⓒ (2 marks)

5 The weight of a bag of flour is 2 kg, but it is found to have a weight of 2.21 kg.
Calculate the percentage error. Ⓒ (2 marks)

flour

....

....

Score   / 17

How well did you do? ✗ 1–6 Try again   7–11 Getting there   12–18 Good work   19–28 Excellent! ✓

For more information on this topic, see pages 28–29 of your Success Guide.

27

Algebra

A

Choose just one answer, a, b, c or d.

1 What is the expression $7a - 4b + 6a - 3b$ when it is fully simplified? *(1 mark)*

  a) $7b - a$ b) $13a + 7b$
  c) $a - 7b$ d) $13a - 7b$

2 If $m = \sqrt{\dfrac{r^2 p}{4}}$ and $r = 3$ and $p = 6$, what is the value of m to 1 decimal place? Ⓒ *(1 mark)*

  a) 13.5 b) 182.3
  c) 3.7 d) 3

3 What is $(n - 3)^2$ when it is multiplied out and simplified? *(1 mark)*

  a) $n^2 + 9$ b) $n^2 + 6n - 9$
  c) $n^2 - 6n - 9$ d) $n^2 - 6n + 9$

4 $P = a^2 + b$. Rearrange this formula to make a the subject. *(1 mark)*

  a) $a = \pm\sqrt{(P - b)}$

  b) $a = \pm\sqrt{(P + b)}$

  c) $a = \dfrac{P - b}{2}$

  d) $a = \dfrac{P + b}{2}$

5 Factorising $n^2 + 7n - 8$ gives: *(1 mark)*

  a) $(n - 2)(n - 6)$
  b) $(n - 2)(n + 4)$
  c) $(n - 1)(n + 8)$
  d) $(n + 1)(n - 8)$

Score   / 5

B

Answer all parts of the questions.

1 John buys b books costing £6 each and p magazines costing 67 pence each. Write down a formula for the total cost (T) of the books and magazines. *(2 marks)*

$T = $ ....

2 $a = \dfrac{b^2 + 2c}{4}$

  a)   Ⓒ Calculate a if $b = 2$ and $c = 6$.

.... *(1 mark)*

  b) Calculate a if $b = 3$ and $c = 5.5$. .... *(1 mark)*

  c) Calculate b if $a = 25$ and $c = 18$. .... *(1 mark)*

3 Factorise the following expressions.

  a) $10n + 15$ .... *(1 mark)*   b) $24 - 36n$ .... *(1 mark)*

  c) $n^2 + 6n + 5$ .... *(1 mark)*   d) $n^2 - 64$ .... *(1 mark)*

  e) $n^2 - 3n - 4$ .... *(1 mark)*

4 Rearrange each of the formulae below to make b the subject.

  a) $p = 3b - 4$ .... *(1 mark)*

  b) $y = \dfrac{b^2 - 6}{4}$ .... *(1 mark)*

  c) $5(n + b) = 2b + 2$ .... *(1 mark)*

Ⓒ *Indicates that a calculator may be used*

These are GCSE-style questions. Answer all parts of the questions. Show your workings (on separate paper if necessary) and include the correct units in your answers.

1 Peter uses this formula to calculate the value of V. Ⓒ (3 marks)

$$V = \frac{\pi x (2R^2 + t^2)}{500}$$

$\pi = 3.14, \quad x = 20, \quad R = 5.2, \quad t = -4.1$

Calculate the value of V, giving your answer to 2 significant figures. Ⓒ

....

....

$V = $ ....

2 a) Expand and simplify $3(2x + 1) - 2(x - 2)$ (2 marks)

....

b) (i) Factorise $6a + 12$ (1 mark)

....

(ii) Factorise completely $10a^2 - 15ab$ (2 marks)

....

c) (i) Factorise $n^2 + 5n + 6$ (2 marks)

....

(ii) Hence simplify fully $\dfrac{2(n + 3)}{n^2 + 5n + 6}$ (2 marks)

....

d) Factorise $(x + y)^2 - 2(x + y)$ (2 marks)

....

3 Show that $(n - 1)^2 + n + (n - 1)$ simplifies to n^2 (3 marks)

....

....

....

4 Simplify fully $\dfrac{x^2 - 8x}{x^2 - 9x + 8}$ (3 marks)

....

Score   / 20

How well did you do?   ✗ 1–13 **Try again**   14–18 **Getting there**   19–30 **Good work**   31–38 **Excellent!** ✓

For more information on this topic, see pages 32–35 of your Success Guide.

29

Equations

A
Choose just one answer, a, b, c or d.

1 Solve the equation $4n - 2 = 10$ (1 mark)

  a) $n = 4$ b) $n = 2$

  c) $n = 3$ d) $n = 3.5$

2 Solve the equation $4(x + 3) = 16$ (1 mark)

  a) $x = 9$ b) $x = 7$

  c) $x = 4$ d) $x = 1$

3 Solve the equation $4(n + 2) = 8(n - 3)$ (1 mark)

  a) $n = 16$ b) $n = 8$

  c) $n = 4$ d) $n = 12$

4 Solve the equation $10 - 6n = 4n - 5$ (1 mark)

  a) $n = 2$

  b) $n = -2$

  c) $n = 1.5$

  d) $n = -1.5$

5 What is the value of k in $y^k = y^{\frac{3}{2}} \div \frac{1}{\sqrt{y}}$ (1 mark)

  a) 2

  b) $\frac{1}{2}$

  c) 1

  d) -2

Score   / 5

B
Answer all parts of the questions.

1 Solve the following equations. (6 marks)

  a) $5n = 25$ ....   b) $\frac{n}{3} = 12$ ....

  c) $2n - 4 = 10$ ....   d) $3 - 2n = 14$ ....

  e) $\frac{n}{5} + 2 = 7$ ....   f) $4 - \frac{n}{2} = 2$ ....

2 Solve the following equations. (4 marks)

  a) $12n + 5 = 3n + 32$ ....   b) $5n - 4 = 3n + 6$ ....

  c) $5(n + 1) = 25$ ....   d) $4(n - 2) = 3(n + 2)$ ....

3 Solve the following equations. (4 marks)

  a) $n^2 - 4n = 0$ ....   b) $n^2 + 6n + 5 = 0$ ....

  c) $n^2 - 5n + 6 = 0$ ....   d) $n^2 - 3n - 28 = 0$ ....

4 The angles in a triangle add up to 180°. Form an equation in n and solve it. (2 marks)

$n = $ ....

Score   / 16

C These are GCSE-style questions. Answer all parts of the questions. Show your workings (on separate paper if necessary) and include the correct units in your answers.

1 Solve these equations.

a) $5m - 3 = 12$ (2 marks)

$m =$ ....

b) $8p + 3 = 9 - 2p$ (2 marks)

$p =$ ....

c) $5(x - 1) = 3x + 7$ (2 marks)

$x =$ ....

d) $\dfrac{w}{2} + \dfrac{(3w + 2)}{3} = \dfrac{1}{3}$ (2 marks)

$w =$ ....

2 a) Factorise $x^2 - 4x + 3$ (2 mark)

....

b) Hence solve the equation $x^2 - 4x + 3 = 0$ (1 mark)

$x =$ ....

and $x =$ ....

3 The area of the rectangle is **125 cm²**. Work out the value of k. (3 marks)

5^{k+1}

$e5$

$k =$ ....

4 Solve the following equations (1 mark)

(i) $3^{2k+1} = 27$

....

(ii) $16^{k-3} = 64$ (2 marks)

....

Score ..... / 17

For more information on this topic, see pages **36–41** of your Success Guide.

31

Equations & inequalities

A Choose just one answer, a, b, c or d.

1 Solve these simultaneous equations to find the values of a and b. **(1 mark)**

$a + b = 10$

$2a - b = 2$

a) $a = 4, b = 6$ b) $a = 4, b = -2$

c) $a = 5, b = 5$ d) $a = 3, b = 7$

2 Solve these simultaneous equations to find the values of x and y. **(1 mark)**

$3x - y = 7$

$2x + y = 3$

a) $x = 3, y = 2$ b) $x = 2, y = 1$

c) $x = -3, y = 2$ d) $x = 2, y = -1$

3 The equation $y^3 + 2y = 82$ has a solution between 4 and 5. By using a method of trial and improvement, find the solution to one decimal place. Ⓒ **(1 mark)**

a) 3.9   b) 4.1   c) 4.2   d) 4.3

4 Solve the inequality $3x + 1 < 19$ **(1 mark)**

a) $x < 3$ b) $x < 7$ c) $x < 5$ d) $x < 6$

5 Solve the inequality $2x - 7 < 9$ **(1 mark)**

a) $x < 9$ b) $x < 10$

c) $x < 8$ d) $x < 6.5$

Score   / 5

B Answer all parts of the questions.

1 Solve these simultaneous equations to find the values of a and b.

a) $2a + b = 8$

$3a - b = 2$ **(2 marks)**

$a = $ ....

$b = $ ....

b) $5a + b = 24$

$2a + 2b = 24$ **(2 marks)**

$a = $ ....

$b = $ ....

c) $a - b = 7$

$3a + 2b = 11$ **(2 marks)**

$a = $ ....

$b = $ ....

d) $4a + 3b = 6$

$2a - 3b = 12$ **(2 marks)**

$a = $ ....

$b = $ ....

2 Use a trial and improvement method to solve the following equation. Give your answer to one decimal place. Ⓒ **(2 marks)**

$t^2 - 2t = 20$ $t = $ ....

3 Solve the following inequalities.

a) $5x + 2 < 12$ **(1 mark)**

....

c) $3 \leq 2x + 1 \leq 9$ **(1 mark)**

....

b) $\frac{x}{3} + 1 \geq 3$ **(1 mark)**

....

d) $3 \leq 3x + 2 \leq 8$ **(1 mark)**

....

Score   / 14

Ⓒ *Indicates that a calculator may be used*

C These are GCSE-style questions. Answer all parts of the questions. Show your workings (on separate paper if necessary) and include the correct units in your answers.

1 n is an integer.

a) Write down the values of n which satisfy the inequality $-4 < n \leq 2$ (2 marks)

....

b) Solve the inequality $5p - 2 \leq 8$ (2 marks)

....

2 Use the method of trial and improvement to solve the equation $x^3 + 3x = 28$
Give your answer correct to one decimal place.
You must show all your working. **C**   (4 marks)

....

....

....

$x = $ ....

3 Solve these simultaneous equations.   (4 marks)

$3x - 2y = -12$
$2x + 6y = 3$

$x = $ ....

$y = $ ....

4 Solve these simultaneous equations.

a) $2a - b = 14$
$a + 3b = 14$ (2 marks)

$a = $ ....

$b = $ ....

b) $5a + 4b = 23$
$3a - 5b = -1$ (2 marks)

$a = $ ....

$b = $ ....

Score   / 16

How well did you do? ✗ 1–9 **Try again**   10–19 **Getting there**   20–27 **Good work**   28–35 **Excellent!** ✓

For more information on this topic, see pages 38–42 of your Success Guide.

33

Further algebra & equations

A Choose just one answer, a, b, c or d.

1 The quadratic equation $x^2 + 4x + 7$ is written in the form $(x + a)^2 + b$. What are the values of a and b? *(1 mark)*

a) $a = 3$, $b = 6$ b) $a = 4$, $b = -3$

c) $a = 2$, $b = 3$ d) $a = 4$, $b = 2$

2 What are the solutions of the quadratic equation $2x^2 + 5x + 2 = 0$? *(1 mark)*

a) $x = -\frac{1}{2}$, $x = -2$ b) $x = \frac{1}{2}$, $x = -2$

c) $x = -\frac{1}{2}$, $x = 2$ d) $x = 2$, $x = -2$

3 What are the solutions of the quadratic equation $x^2 - 4 = 0$? *(1 mark)*

a) $x = 2$, $x = 2$ b) $x = -2$, $x = -2$

c) $x = 0$, $x = 4$ d) $x = 2$, $x = -2$

4 What are the solutions of the quadratic equation $6x^2 + 2x = 8$? *(1 mark)*

a) $x = 1$, $x = \frac{3}{4}$ b) $x = 1$, $x = -\frac{4}{3}$

c) $x = -1$, $x = \frac{4}{3}$ d) $x = -1$, $x = -\frac{3}{4}$

5 The quadratic equation $x^2 - 2x + 3$ is written in the form $(x + a)^2 + b$. What are the values of a and b? *(1 mark)*

a) $a = 1$, $b = 2$ b) $a = -1$, $b = -2$

c) $a = -1$, $b = 2$ d) $a = 1$, $b = -2$

Score   / 5

B Answer all parts of the questions.

1 a) Factorise $x^2 + 11x + 30$ .... *(2 marks)*

b) Write the following as a single fraction in its simplest form: *(4 marks)*

$$\frac{4}{x+6} + \frac{4}{x^2+11x+30}$$

....

2 $(x + 4)(x - 3) = 2$ Ⓒ

a) Show that $x^2 + x - 14 = 0$ *(2 marks)*

....

b) Solve the equation $x^2 + x - 14 = 0$
Give your answers correct to 3 significant figures. *(3 marks)*

....

3 The formula $a = \dfrac{3(b + c)}{bc}$ is rearranged to make c the subject. Greg says the answer is $c = \dfrac{3b}{ab - 3}$.
Decide whether Greg is right. You must justify your answer. *(3 marks)*

....

....

Score   / 14

Ⓒ *Indicates that a calculator may be used*

C These are GCSE-style questions. Answer all parts of the questions. Show your workings (on separate paper if necessary) and include the correct units in your answers.

1 The diagram shows a right-angled triangle with base $(x - 3)$ and height $(x + 4)$.
All measurements are given in centimetres.

The area of the triangle is 12 square centimetres.

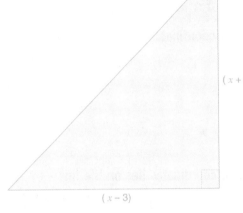

$(x + 4)$

$(x - 3)$

a) Show that $x^2 + x - 36 = 0$ (C)   (3 marks)

....

....

b) Find the length of the base of the triangle. Give your answer correct to 2 d.p. (C)   (4 marks)

....

....

.... cm

2 Make b the subject of the formula $a = \dfrac{8b+5}{4 - 3b}$ (4 marks)

....

....

3 The expression $x^2 + 6x + 3$ can be written in the form $(x + a)^2 + b$ for all values of x.   (3 marks)

a) Find a and b.

$a = $ ....

$b = $ ....

b) The expression $x^2 + 6x + 3$ has a minimum value. Find this minimum value.   (1 mark)

....

....

4 Make a the subject of the formula $P = 4a + \pi a + 3b$ (3 marks)

....

....

Score   / 18

How well did you do? ✗ 1–7 **Try again**   8–18 **Getting there**   19–28 **Good work**   29–37 **Excellent!** ✓

For more information on this topic, see pages 35–41 of your Success Guide.

35

Straight line graphs

A Choose just one answer, a, b, c or d.

1 Which pair of coordinates lies on the line $x = 2$? *(1 mark)*

a) (1, 3)
b) (2, 3)
c) (3, 2)
d) (0, 2)

2 Which pair of coordinates lies on the line $y = -3$? *(1 mark)*

a) (−3, 5)
b) (5, −2)
c) (−2, 5)
d) (5, −3)

3 What is the gradient of the line $y = 2 - 5x$? *(1 mark)*

a) −2   b) −5   c) 2   d) 5

4 These graphs have been drawn: $y = 3x - 1$, $y = 5 - 2x$, $y = 6x + 1$, $y = 2x - 3$
Which graph is the steepest? *(1 mark)*

a) $y = 3x - 1$
b) $y = 5 - 2x$
c) $y = 6x + 1$
d) $y = 2x - 3$

5 At what point does the graph $y = 3x - 4$ intercept the y axis? *(1 mark)*

a) (0, −4)
b) (0, 3)
c) (−4, 0)
d) (3, 0)

Score   / 5

B Answer all parts of the questions.

1 a) On the grid, draw the graph of $y = 6 - x$.
Join your points with a straight line. *(2 marks)*

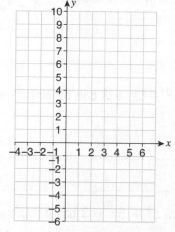

b) A second line goes through the coordinates (1, 5), (−2, −4) and (2, 8).
i) Draw this line on the grid. *(1 mark)*
ii) Write down the equation of the line you have just drawn. *(2 marks)*

....

c) What are the coordinates of the point where the two lines meet? *(1 mark)*

....

2 The equations of five straight lines are: $y = 2x - 4$, $y = 3 - 2x$, $y = 4 - 2x$, $y = 5x - 4$, $y = 3x - 5$
Two of the lines are parallel. Write down the equations of these two lines.

.... and .... *(2 marks)*

Score   / 8

C

These are GCSE-style questions. Answer all parts of the questions. Show your workings (on separate paper if necessary) and include the correct units in your answers.

1 The line with equation $3y + 2x = 12$ has been drawn on the grid.

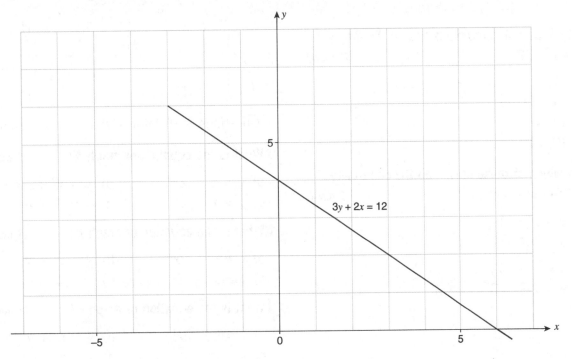

a) Write down the gradient of the line $3y + 2x = 12$ (2 marks)

....

b) Write down the equation of the line which is parallel to $3y + 2x = 12$ and passes through the point with coordinates $(0, 2)$. (1 mark)

....

c) On the grid above, draw the graph with the equation $3y - x = 3$ (2 marks)

d) Write down the coordinates of the point of intersection of the two straight-line graphs. (1 mark)

(..............., ...)

e) Write down the gradient of a line that is perpendicular to the line $3y + 2x = 12$ (1 mark)

....

Score   / 7

How well did you do?   ✗ 1–4 **Try again**   5–9 **Getting there**   10–14 **Good work**   15–20 **Excellent!** ✓

For more information on this topic, see pages 44–45 of your Success Guide.

37

Curved graphs

A Choose just one answer, a, b, c or d.

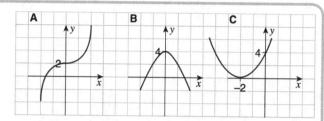

A   **B**   **C**

Questions 3–5 refer to the diagrams above.

1 Which pair of coordinates lies on the graph
$y = x^2 - 2$? **(1 mark)**

a) (1, 1)
b) (4, 14)
c) (2, 4)
d) (0, 2)

2 On which of these curves do the coordinates
(2, 5) lie? **(1 mark)**

a) $y = x^2 - 4$
b) $y = 2x^2 + 3$
c) $y = x^2 - 6$
d) $y = 2x^2 - 3$

3 What is the equation of graph A? **(1 mark)**

a) $y = 5 - 2x^2$ b) $y = x^2 + 4x + 4$
c) $y = x^3 + 2$ d) $y = 4 - x^2$

4 What is the equation of graph B? **(1 mark)**

a) $y = 5 - 2x^2$ b) $y = x^2 + 4x + 4$
c) $y = x^3 + 2$ d) $y = 4 - x^2$

5 What is the equation of graph C? **(1 mark)**

a) $y = 5 - 2x^2$ b) $y = x^2 + 4x + 4$
c) $y = x^3 + 2$ d) $y = 4 - x^2$

Score   / 5

B Answer all parts of the questions.

1 a) Complete the table of values for $y = x^2 - 2x - 2$ **(2 marks)**

x	−2	−1	0	1	2	3
$y = x^2 - 2x - 2$			−2			1

b) On the grid below, draw the graph of $y = x^2 - 2x - 2$ **(3 marks)**

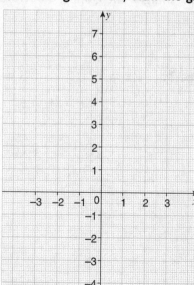

c) Use your graph to write down an estimate for:

i) the minimum value of y.

$y =$............................ **(1 mark)**

ii) the solutions of the equation $x^2 - 2x - 2 = 0$.

$x =$......................... and $x =$......................... **(2 marks)**

Score   / 8

C These are GCSE-style questions. Answer all parts of the questions. Show your workings (on separate paper if necessary) and include the correct units in your answers.

1 a) Complete the table of values for the graph of $y = x^3 - 4$ (2 marks)

x	-2	-1	0	1	2	3
$y = x^3 - 4$		-5				23

b) On the grid, draw the graph of $y = x^3 - 4$ for values of x between -2 and 3. (2 marks)

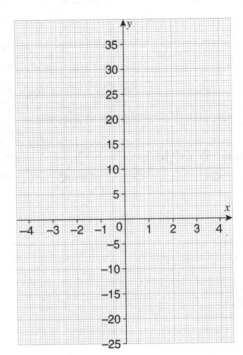

c) Use your graph to find an estimate of:

i) the solution of the equation $x^3 - 4 = 0$ (1 mark)

$x =$ ....

ii) the solution of the equation $x^3 - 4 = 10$ (2 marks)

$x =$ ....

iii) the solution of the equation $x^3 - 4 = 2$ (2 marks)

$x =$ ....

iv) the solution of the equation $x^3 - 4 = -7$ (2 marks)

$x =$ ....

Score   / 11

How well did you do? ✗ 1–4 Try again   5–9 Getting there   10–16 Good work   17–24 Excellent! ✓

For more information on this topic, see pages 46–47 of your Success Guide.

39

Harder work on graphs

A Choose just one answer, a, b, c or d.

1 Solve these simultaneous equations
to find values for a and b.   **(1 mark)**
$a^2 + b^2 = 13$
$2a + b = 7$

a) $a = -2, b = 3$
b) $a = 3, b = 2$
c) $a = 2, b = 3$
d) $a = 2, b = -3$

2 Solve these simultaneous equations
to find values for a and b.   **(1 mark)**
$a^2 - b = 3$
$3a + b = 1$

a) $a = 1, b = -2$
b) $a = -1, b = 2$
c) $a = -1, b = -2$
d) $a = 1, b = 2$

3 What are the coordinates of a point
where the line $y = 4 - x$ and the
circle $x^2 + y^2 = 40$ meet?   **(1 mark)**

a) $x = 2, y = -6$
b) $x = 2, y = 6$
c) $x = -2, y = -6$
d) $x = -2, y = 6$

4 The graph $y = x^2$ is translated 2 units
to the left. What is the equation of the
new curve?   **(1 mark)**

a) $y = (x - 2)^2$
b) $y = x^2 + 2$
c) $y = (x + 2)^2$
d) $y = x^2 - 2$

Score   / 4

B Answer all parts of the questions.

1 This is a sketch of the curve with equation $y = f(x)$.
The maximum point of the curve is A (2, 7).
Write down the coordinates of the maximum point
of each of the following curves.

a) $y = f(x) - 3$ (............, ...)
b) $y = f(x + 1)$ (............, ...)
c) $y = f(x - 4)$ (............, ...)
d) $y = f(-x)$ (............, ...)
e) $y = f(2x)$ (............, ...)

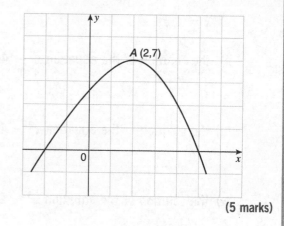

(5 marks)

2 Decide whether this statement is true or false:
'The line $2x - y = 6$ intersects with the circle $x^2 + y^2 = 17$ at the point (1, −4).'
Explain your reasoning.

   (2 marks)

....

....

Score   / 7

40

C These are GCSE-style questions. Answer all parts of the questions. Show your workings (on separate paper if necessary) and include the correct units in your answers.

1 a) Solve these simultaneous equations.   (5 marks)

$x + 3y = -12$

$x^2 + y^2 = 34$

....

....

b) Give a geometrical interpretation of the result.   (2 marks)

....

....

2 The graph of $y = f(x)$ is sketched in the diagrams below.

a) Sketch the graph of $y = f(-x)$ on this grid.   (2 marks)

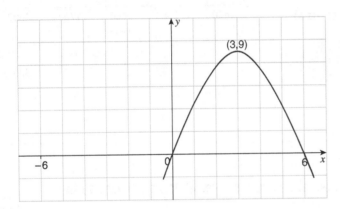

b) Sketch the graph of $y = f(x + 2)$ on this grid.   (2 marks)

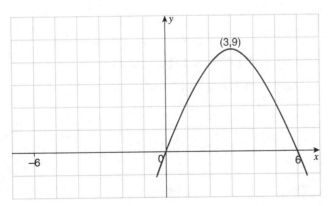

Score   / 11

For more information on this topic, see pages 48–49 of your Success Guide.

Interpreting graphs

A   Choose just one answer, a, b, c or d.

Use the graph opposite for these questions.
The graph represents Mrs Morgan's car journey.

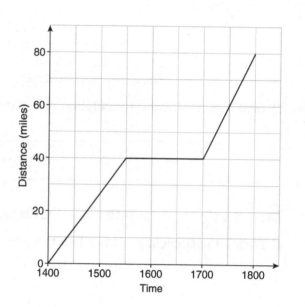

1 At what speed did Mrs Morgan travel for the first hour and a half? **(1 mark)**

a) 25 mph   b) 28 mph
c) 30 mph   d) 26.7 mph

2 At what time did Mrs Morgan take a break from her car journey? **(1 mark)**

a) 1530   b) 1600
c) 1400   d) 1500

3 At what speed did Mrs Morgan travel between 1700 and 1800 hours? **(1 mark)**

a) 60 mph   b) 80 mph
c) 35 mph   d) 40 mph

Score   / 3

B   Answer all parts of the questions.

1 Water is poured into these odd-shaped vases at a constant rate. Match each vase to the correct graph. **(3 marks)**

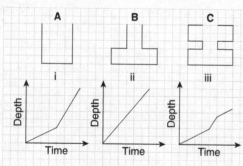

Vase A matches graph ....
Vase B matches graph ....
Vase C matches graph ....

2 This is the graph of $y = x^2 - x - 6$

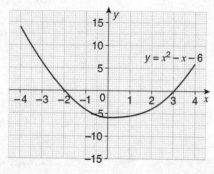

a) Use the graph to find the roots of the equation
$x^2 - x - 6 = 0$ **(2 marks)**

.... and ....

b) By drawing suitable straight lines on the graph, solve these equations:

i) $x^2 - x - 6 = 2$ .... **(2 marks)**

ii) $x^2 - 7 = 0$ .... **(3 marks)**

Score   / 10

C These are GCSE-style questions. Answer all parts of the questions. Show your workings (on separate paper if necessary) and include the correct units in your answers.

1 In an experiment, these values of the variables R and T were obtained.

R	0	1	2	3	4
T	5	8	17	32	53

The variables R and T are thought to satisfy a relationship of the form $T = aR^2 + b$

a) On the axes, draw a graph to test this.   (4 marks)

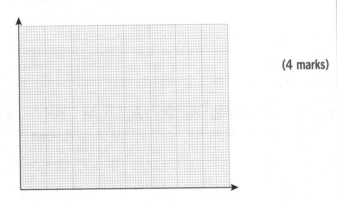

b) Use your graph to estimate the values of a and b.   (2 marks)

$a = $ ....

$b = $ ....

c) Use your relationship to find T when $R = 12$.   (1 mark)

$T = $ ....

2 The sketch graph shows a curve $y = ab^x$ where $b > 0$.

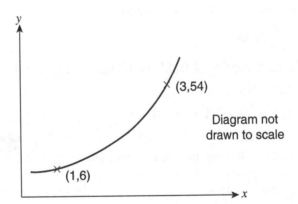

(3,54)

Diagram not drawn to scale

(1,6)

The curve passes through the points (1, 6) and (3, 54).
Calculate the values of a and b.   (3 marks)

$a = $ ....

$b = $ ....

Score   / 10

How well did you do? ✗ 1–4 Try again   5–9 Getting there   10–16 Good work   17–23 Excellent! ✓

For more information on this topic, see pages 50–51 of your Success Guide.

43

Bearings and scale drawings

A Choose just one answer, a, b, c or d.

1 The bearing of P from Q is 050°. What is the bearing of Q from P? *(1 mark)*

a) 130°
b) 50°
c) 230°
d) 310°

2 The bearing of R from S is 130°. What is the bearing of S from R? *(1 mark)*

a) 310°
b) 230°
c) 050°
d) 200°

3 The bearing of A from B is 240°. What is the bearing of B from A? *(1 mark)*

a) 120°
b) 60°
c) 320°
d) 060°

4 The length of a car park is 25 metres. A scale diagram of the car park is being drawn to a scale of 1 cm to 5 metres. What is the length of the car park on the scale diagram? *(1 mark)*

a) 500 mm
b) 5 cm
c) 50 cm
d) 5 m

Score / 4

B Answer all parts of the questions.

1 The scale on a road map is 1 : 50 000. Two towns are 20 cm apart on the map. Work out the real distance, in km, between the two towns. *(2 marks)*

.... km

2 A ship sails on a bearing of 065° for 10 km. It then continues on a bearing of 120° for a further 15 km to a port (P).

a) On a separate piece of paper, draw, using a scale of 1 cm to 2 km, an accurate scale drawing of this information. *(3 marks)*

b) Measure on your diagram the direct distance between the starting point and the port (P). *(1 mark)*

.... km

c) What is the bearing of port P from the starting point? *(1 mark)*

.... °

3 Is this statement true or false? *(1 mark)*

'The bearing of B from A is 060°.'

....

Score / 8

C   These are GCSE-style questions. Answer all parts of the questions. Show your workings (on separate paper if necessary) and include the correct units in your answers.

1 The scale drawing shows the positions of points A, B, C and D. Point C is due east of point A.

Scale: 1cm represents 50m

a) Use measurements from the drawing to find:

   i) the distance, in metres, of B from A.   .... m   (1 mark)

   ii) the bearing of B from A.   .... °   (2 marks)

   iii) the bearing of D from B.   .... °   (2 marks)

b) Point E is 250 m from point C on a bearing of 055°. Mark the position of point E on the diagram above.   (2 marks)

2 Here is a sketch of a triangle. Use a scale of 1 cm to 2 m to make an accurate scale drawing of the triangle.   (3 marks)

Diagram not drawn to scale

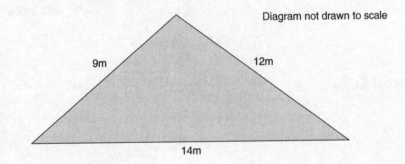

Score   / 10

How well did you do?   ✗ 1–4 Try again   5–10 Getting there   11–16 Good work   17–22 Excellent! ✓

For more information on this topic, see pages 58–59 of your Success Guide.

45

Transformations 1

A Choose just one answer, a, b, c or d.

Questions 1–4 refer to the diagram opposite.

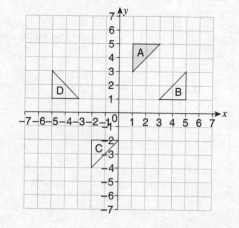

1 What is the transformation that would map
   shape A onto shape B?   (1 mark)

   a) reflection   b) rotation
   c) translation   d) enlargement

2 What is the transformation that would map
   shape A onto shape C?   (1 mark)

   a) reflection   b) rotation
   c) translation   d) enlargement

3 What is the transformation that would map
   shape A onto shape D?   (1 mark)

   a) reflection   b) rotation
   c) translation   d) enlargement

4 What special name is given to the relationship
   between triangles A, B, C, and D?   (1 mark)

   a) enlargement   b) congruent
   c) translation   d) similar

Score   / 4

B Answer all parts of the questions.

1 On the grid, carry out the following
   transformations.   (3 marks)

   a) Reflect shape A in the y axis.
      Call the new shape R.

   b) Rotate shape A 90° clockwise, about (0, 0).
      Call the new shape S.

   c) Translate shape A by the vector $\begin{pmatrix} -3 \\ 4 \end{pmatrix}$.
      Call the new shape T.

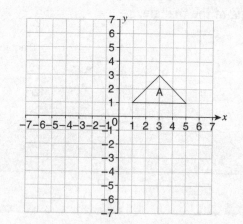

2 All the shapes in the diagram are either a
   reflection, rotation or translation of object P.
   State the transformation that has taken place
   in each of the following.

   a) P is transformed to A ....

   b) P is transformed to B ....

   c) P is transformed to C ....

   d) P is transformed to D ....

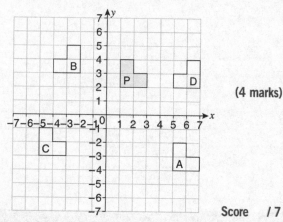

(4 marks)

Score   / 7

46

C These are GCSE-style questions. Answer all parts of the questions. Show your workings (on separate paper if necessary) and include the correct units in your answers.

1

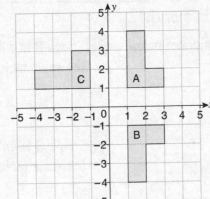

a) Describe fully the single transformation which takes shape A onto shape B.   (2 marks)

....

....

b) Describe fully the single transformation which takes shape A onto shape C.   (3 marks)

....

....

2

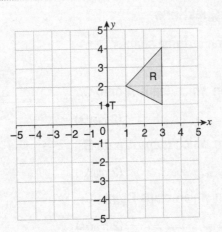

The triangle R has been drawn on the grid.

a) Rotate triangle R 90° clockwise about the point T (0, 1) and call the image P.   (3 marks)

b) Translate triangle R by the vector $\begin{pmatrix} -4 \\ -3 \end{pmatrix}$ and call the image Q.   (3 marks)

Score   / 11

How well did you do?   ✗ 1–6 Try again   7–10 Getting there   11–16 Good work   17–22 Excellent! ✓

For more information on this topic, see pages 60–61 of your Success Guide.

4

Transformations 2

A

Choose just one answer, a, b, c or d.

Questions 1–3 refer to the diagram drawn below.

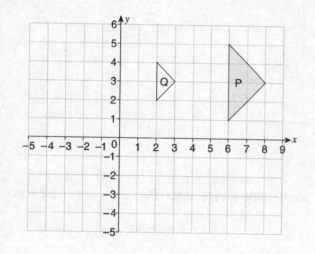

1 Shape P is enlarged to give shape Q. What is the scale factor of the enlargement? **(1 mark)**

a) $\frac{1}{3}$ b) 2

c) 3   d) $\frac{1}{2}$

2 Shape Q is enlarged to give shape P. What is the scale factor of the enlargement? **(1 mark)**

a) $\frac{1}{3}$ b) 2

c) 3   d) $\frac{1}{2}$

3 What are the coordinates of the centre of enlargement in both cases? **(1 mark)**

a) $(3, -2)$ b) $(-2, 3)$

c) $(-3, 4)$ d) $(0, 0)$

Score   / 3

B

Answer all parts of the questions.

1 The diagram shows the position of three shapes, A, B and C.

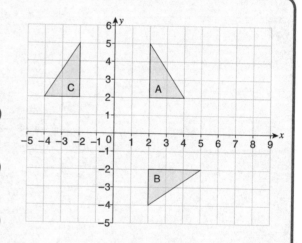

a) Describe the transformation which maps A onto C.

.... **(2 marks)**

b) Describe the transformation which maps A onto B.

.... **(2 marks)**

c) Describe the transformation which maps B onto C.

.... **(2 marks)**

2 On the grid, enlarge triangle PQR by a scale factor of −2 with centre of enlargement (0, 0), and call the image P'Q'R'.

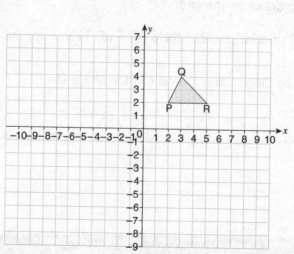

(3 marks)

Score   / 9

48

GCSE Success

Workbook Answer Booklet

Mathematics Higher

Fiona C. Mapp

Answers

NUMBER

Fractions

A
1. c
2. a
3. b
4. c
5. a

B
1. a) $\frac{5}{9}$
   b) $\frac{17}{44}$
   c) $\frac{3}{14}$
   d) 3
   e) $1\frac{2}{3}$
   f) $\frac{5}{9}$
   g) $\frac{7}{8}$
   h) $2\frac{34}{49}$
2. a) $\frac{1}{7}$ $\frac{3}{10}$ $\frac{1}{2}$ $\frac{2}{3}$ $\frac{3}{4}$ $\frac{4}{5}$
   b) $\frac{1}{9}$ $\frac{2}{7}$ $\frac{1}{3}$ $\frac{2}{5}$ $\frac{5}{8}$ $\frac{3}{4}$
3. a) true
   b) false
4. 28 students

C
1. Jonathan, since $\frac{2}{3}$ is greater than $\frac{3}{5}$.
2. a) $1\frac{7}{15}$
   b) $1\frac{16}{33}$
   c) $\frac{8}{63}$
   d) $\frac{3}{4}$
3. $\frac{7}{15}$

Percentages 1

A
1. c
2. c
3. a
4. d
5. b

B
1. a) £12
   b) £45
   c) £4
   d) 5 g
2. £26 265
3. 39.3% (3s.f.)
4. 77%
5. £20
6. £225
7. 40%

C
1. a) £30.63
   b) £339.15
2. 20%
3. £70

Percentages 2

A
1. d
2. a
3. c
4. b

B
1. £168.03
2. £75.60
3. £6 969.60
4. £515.88
5. £135 475.20
6. 80.2 pence

C
1. a) £1000   b) £160
2. £1 337.11
3. a) 3.5% b) £757.12
4. a) 4.5% b) £59 728.23

Fractions, decimals and percentages

A
1. c
2. b
3. d
4. d
5. a

B
1.

Fraction	Decimal	Percentage
$\frac{2}{5}$	0.4	40%
$\frac{1}{20}$	0.05	5%
$\frac{1}{3}$	0.3	33.$\dot{3}$%
$\frac{1}{25}$	0.04	4%
$\frac{1}{4}$	0.25	25%
$\frac{1}{8}$	0.125	12.5%

2. 30% $\frac{1}{3}$ 0.37 $\frac{3}{8}$ $\frac{1}{2}$ 0.62   92%
3. Both will give the same answer because increasing by 20% is the same as multiplying by 1.2. Finding 10% then doubling it gives 20%, which when you add it to 40, is the same as increasing £40 by 20%.

C
1. $\frac{1}{8}$ 25%   0.27 $\frac{1}{3}$ $\frac{2}{5}$ 0.571   72%
2. a) Ed's Electricals   : £225
   Sheila's Bargains : £217.38
   Gita's TV shop   : £232
   Maximum price = £232
   Minimum price = £217.38
   b) £200
3. 'Rosebushes' is cheaper because $\frac{1}{4}$ = 25% which is greater than the offer at 'Gardens are Us'.

Approximations and using a calculator

A
1. b
2. d
3. c
4. c
5. d

B
1. a) true
   b) false
   c) false
   d) true
2. a) 100
   b) 90
3. a) 365 (3 s.f.)
   b) 10.2 (3 s.f.)
c) 6320 (3 s.f.)
d) 0.0812 (3 s.f.)

C
1. a) 30 and 80
   b) 2400
   c) 49
2. 7.09 (3s.f.)
3. a) 5.937 102 5
   b) $\frac{30 \times 6}{40 - 10} = \frac{180}{30} = 6$
4. a) B
   b) B
   c) B
   d) A

Ratio

A
1. c
2. d
3. c
4. d
5. b

B
1. a) 1 : 1.5
   b) 1 : 1.$\dot{6}$
   c) 1 : 3
2. 1200 ml
3. a) £10.25
   b) 48 g
4. £25 000
5. 4.5 days

C
1. Vicky £6 400
   Tracey £8 000
2. butter   75 g
   sugar   60 g
   eggs   3
   flour   67.5 g
   milk   22.5 ml
3. 6 days

Indices

A
1. a
2. c
3. d
4. a
5. b

B
1. a) false
   b) false
   c) true
   d) true
   e) false
   f) true
2. a) 1
   b) $16a^8$
   c) $\frac{3}{4}a^{-3}$
   d) $27a^6b^9$
3. a) $4x^{-2}$
   b) a^2b^{-3}
   c) $3y^{-5}$
4. a) $\pm\frac{1}{5}$
   b) 343
   c) $\frac{25}{16}$
   d) $\frac{1}{27}$

C
1. a) p^7
   b) n^{-4} or $\frac{1}{n^4}$
c) a^6
d) $4ab$
2. a) 1
   b) $\frac{1}{81}$
   c) 648
   d) 16
   e) $\frac{1}{5}$
3. a) i) 1
   ii) $\frac{1}{16}$
   iii) $\pm\frac{3}{2} = \pm 1\frac{1}{2}$
   b) 5^4
4. a) $\frac{1}{125}$
   b) $\frac{9}{4}$
   c) $\frac{1}{4}$
5. a) 2^{-1}
   b) 2^{20}
   c) $2^{\frac{5}{2}}$
6. a) $\frac{1}{125x^3}$ b) $\frac{1}{16x^8 y^{12}}$ c) $\frac{1}{27y^3}$
   d) $32x^5 y^{15}$.

Standard index form

A
1. b
2. b
3. a
4. b
5. b

B
1. a) true
   b) false
   c) false
   d) false
2. a) 8×10^{15}
   b) 1.4×10^4
   c) 3×10^{16}
   d) 8×10^3
3. a) 8.19×10^3
   b) 7.56×10^3
   c) 6×10^{-4}
   d) 2.225×10^8
4. 1.8×10^{-7} grams

C
1. a) i) 207 000
   ii) 4.6×10^{-5}
   b) 3.5×10^{12}
2. 2.2×10^{12}
3. 2.6×10^{-6}
4. 1.25×10^{-10}

Recurring decimals and surds

A
1. c
2. d
3. a
4. b
5. a

B
1. 0.$\dot{3}$ $\frac{7}{9}$
   0.$\dot{7}$ $\frac{244}{333}$
   0.2$\dot{4}$ $\frac{13}{30}$
   0.7$\dot{3}\dot{2}$ $\frac{1}{3}$
   0.4$\dot{3}$ $\frac{8}{33}$
2. $\frac{124}{990} = \frac{62}{495}$
3. a) $2\sqrt{6}$
   b) $5\sqrt{3}$
   c) $6\sqrt{3}$
   d) $6\sqrt{5}$

C

1. a) $\frac{6}{11}$

   b) $\frac{26}{990} = \frac{13}{495}$

2. a) 9

   b) $a = 4$

   c) $\frac{3}{5}$

3. $\dfrac{(5 + \sqrt{5})(2 - 2\sqrt{5})}{\sqrt{45}}$

   $= \dfrac{10 - 10\sqrt{5} + 2\sqrt{5} - 2(\sqrt{5})^2}{3\sqrt{5}}$

   $= \dfrac{-8\sqrt{5}}{3\sqrt{5}}$

   $= \dfrac{-8}{3}$

4. $\frac{61}{495}$

5. $19 - 8\sqrt{3}$

6. $\dfrac{\sqrt{125} + \sqrt{50}}{\sqrt{5}}$

   $= \dfrac{5\sqrt{5} + 5\sqrt{2}}{\sqrt{5}}$

   $= \dfrac{(5\sqrt{5} + 5\sqrt{2})}{\sqrt{5}} \times \dfrac{\sqrt{5}}{\sqrt{5}}$

   $= \dfrac{5(\sqrt{5})^2 + 5\sqrt{10}}{5}$

   $= 5 + \sqrt{10}$

7. $1 + \sqrt{2}$

8. $\frac{25}{33}$

9. $0.4\dot{5} = \frac{5}{11}$

   $x = 0.454545. \ldots$

   $100x = 45.454545. \ldots$

   $99x = 45$

   $x = \frac{45}{99}$

   $x = \frac{5}{11}$

Direct and inverse proportion

A

1. d
2. a
3. b
4. c

B

1.

a	10	12	30
b	2	2.4	6

2.

x	2	4	3	5
y	12	48	27	75

3. a) $z = \frac{k}{v^2}$

   b) i) 62.5

      ii) $\frac{1}{2}$

C

1. $E = kF$

   $6 = k \times 15$

   $\therefore k = \frac{2}{5}$

   $E = \frac{2}{5}F$

   When $F = 80$ N, $E = 32$ cm

2. $V = kh^3$

   $60 = k \times 8$

   $k = 7.5$

   $V = 7.5h^3$

   $V = 937.5$ cm³

3. $I = \frac{k}{d^2}$

   $50 = \frac{k}{4}$

   $\therefore k = 200$

   $I = \frac{200}{d^2}$

   a) $I = 16.3$ (1 d.p.)

   b) $d = 4$

4. $c = \frac{k}{b}$

   $\therefore 4 = \frac{k}{10}$ so $k = 40$

   Daisy is correct.

Upper and lower bounds of measurement

A

1. b
2. c
3. c
4. b
5. d

B

1. a) 5.235

   b) 4.5

   c) 0.3245

   d) 8.44

2. Lower bound e

   Upper bound a

C

1. a) 7.9 cm

   b) 0.642 cm

2. Lower bound 3.9717 cm (5 s.f.)

   Upper bound 3.9738 cm (5 s.f.)

3. 0.1811 g cm⁻³ ≤ density < 0.2 g cm⁻³

4. 11.25 cm²

5. 10.5%

ALGEBRA

Algebra

A

1. d
2. c
3. d
4. a
5. c

B

1. $T = 6b + 0.67p$

2. a) 4

   b) 5

   c) 8

3. a) $5(2n + 3)$

   b) $12(2 - 3n)$

   c) $(n + 1)(n + 5)$

   d) $(n - 8)(n + 8)$

   e) $(n + 1)(n - 4)$

4. a) $b = \frac{p + 4}{3}$

   b) $b = \pm\sqrt{(4y + 6)}$

   c) $b = \frac{2 - 5n}{3}$

C

1. $V = 8.9$ (2 s.f.)

2. a) $4x + 7$

   b) i) $6(a + 2)$

      ii) $5a(2a - 3b)$

   c) i) $(n + 2)(n + 3)$

      ii) $\frac{2}{n + 2}$

   d) $(x + y)(x + y + 2)$

3. $(n - 1)^2 + n + (n - 1)$

   $n^2 - 2n + 1 + n + n - 1$

   $= n^2 - 2n + 1 + 2n - 1$

   $= n^2$

4. $\dfrac{x^2 - 8x}{x^2 - 9x + 8} = \dfrac{x(x - 8x)}{(x - 8)(x - 1)}$

   $= \dfrac{x}{(x - 1)}$

Equations

A

1. c
2. d
3. b
4. c
5. a

B

1. a) $n = 5$

   b) $n = 36$

   c) $n = 7$

   d) $n = -5.5$

---

e) $n = 25$

f) $n = 4$

2. a) $n = 3$

   b) $n = 5$

   c) $n = 4$

   d) $n = 14$

3. a) $n = 0, n = 4$

   b) $n = -5, n = -1$

   c) $n = 3, n = 2$

   d) $n = -4, n = 7$

4. $2n + (n + 30°) + (n - 10°) = 180°$

   $4n + 20° = 180°$

   $n = 40°$

C

1. a) $m = 3$

   b) $p = \frac{6}{10}$ or $p = \frac{3}{5}$

   c) $x = 6$

   d) $\dfrac{3w + 2(3w + 2)}{6} = \dfrac{1}{3}$

   $3w + 6w + 4 = \frac{6}{3}$

   $3w + 6w + 4 = 2$

   $9w + 4 = 2$

   $w = -\frac{2}{9}$

2. a) $(x - 1)(x - 3)$

   b) $x = 1$ and $x = 3$

3. $k = \frac{3}{2}$ or 1.5

4. i) $k = 1$

   ii) $k = 4.5$

Equations and inequalities

A

1. a
2. d
3. c
4. d
5. c

B

1. a) $a = 2, b = 4$

   b) $a = 3, b = 9$

   c) $a = 5, b = -2$

   d) $a = 3, b = -2$

2. 5.6 and −3.6

3. a) $x < 2$

   b) $x \geq 6$

   c) $1 \leq x \leq 4$

   d) $\frac{1}{3} < x \leq 2$

C

1. a) $-3 \quad -2 \quad -1 \quad 0 \quad 1 \quad 2$

   b) $p \leq 2$

2. $x = 2.7$

3. $x = -3, y = 1.5$

4. a) $a = 8, b = 2$

   b) $a = 3, b = 2$

Further algebra and equations

A

1. c
2. a
3. d
4. b
5. c

B

1. a) $(x + 6)(x + 5)$

   b) $\dfrac{4}{(x + 6)} + \dfrac{4}{(x^2 + 11x + 30)}$

   $= \dfrac{4}{(x + 6)} + \dfrac{4}{(x + 6)(x + 5)}$

   $= \dfrac{4(x + 5) + 4}{(x + 6)(x + 5)}$

   $= \dfrac{4x + 24}{(x + 6)(x + 5)}$

   $= \dfrac{4}{(x + 5)}$

2. a) $(x + 4)(x - 3) = 2$

   $x^2 + x - 12 = 2$

   $x^2 + x - 14 = 0$

   b) $x = 3.27$ or $x = -4.27$

---

3. $a = \dfrac{3(b + c)}{bc}$

   $abc = 3b + 3c$

   $abc - 3c = 3b$

   $c(ab - 3) = 3b$

   $\therefore c = \dfrac{3b}{ab - 3}$

   Greg is right.

C

1. a) $\frac{1}{2} \times (x - 3) \times (x + 4) = 12$

   $(x - 3)(x + 4) = 24$

   $x^2 + x - 12 = 24$

   $x^2 + x - 36 = 0$

   b) $x = 5.52$

   ∴ Base of triangle is 2.52 cm

2. $a = \dfrac{8b + 5}{(4 - 3b)}$

   $a(4 - 3b) = 8b + 5$

   $4a - 3ab = 8b + 5$

   $4a - 5 = 8b + 3ab$

   $4a - 5 = b(8 + 3a)$

   $b = \dfrac{4a - 5}{8 + 3a}$

3. a) $a = 3, b = -6$

   b) Minimum value is −6.

4. $P = 4a + \pi a + 3b$

   $P - 3b = 4a + \pi a$

   $P - 3b = a(4 + \pi)$

   $a = \dfrac{P - 3b}{4 + \pi}$

Straight line graphs

A

1. b
2. d
3. b
4. c
5. a

B

1. a), b) i)

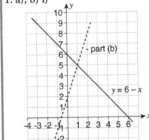

   b) ii) $y = 3x + 2$

   c) $(1, 5)$

2. $y = 3 - 2x$ and $y = 4 - 2x$

C

1. a) Gradient $= -\frac{2}{3}$

   b) $3y + 2x = 6$ or $y = -\frac{2}{3}x + 2$

   c)

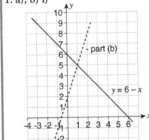

   d) $(3, 2)$

   e) $\frac{3}{2}$

Curved graphs

A

1. b
2. d
3. c
4. d
5. b

B
1. a)

x	-2	-1	0	1	2	3
y	6	1	-2	-3	-2	1

b)

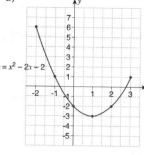

c) i) $y = -3$
   ii) $x = 2.7$, $x = -0.62$

C
1. a)

x	-2	-1	0	1	2	3
y	-12	-5	-4	-3	4	23

b)

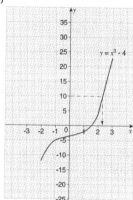

c) i) $x = 1.6$ ii) $x = 2.4$
   iii) $x = 1.8$ iv) $x = -1.4$

Harder work on graphs
A
1. c
2. a
3. d
4. c

B
1. a) (2, 4)
   b) (1, 7)
   c) (6, 7)
   d) (-2, 7)
   e) (1, 7)
2. Statement is true since $x = 1$, $y = -4$ is a simultaneous solution of the two equations:
   i.e. $2 \times 1 - (-4) = 6$
   $1^2 + (-4)^2 = 17$

C
1. a) $x = 3$, $y = -5$ and $x = -5\frac{2}{5}$, $y = -2\frac{1}{5}$
   b) They are the coordinates of the points where the line $3y = -12 - x$ intersects with the circle $x^2 + y^2 = 34$.
2. a)

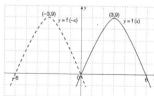

---

b)

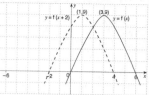

Interpreting graphs
A
1. d
2. a
3. d

B
1. Vase A – graph ii.
   Vase B – graph i.
   Vase C – graph iii.
2. a) Roots are $x = 3$ and $x = -2$ (read where curve crosses x axis)
   b) i) Approximately, $x = 3.3$ and $x = -2.3$ (read across where y = 2)
   ii) Approximately, $x = -2.7$ and $x = 2.6$ (draw the line $y = 1 - x$ and find the point of intersection with the curve)

C
1. a)

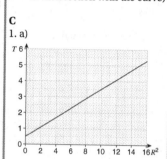

   b) $a = 3$, $b = 5$
   c) approximately 437
2. $a = 2$, $b = 3$

SHAPE, SPACE AND MEASURES
Bearings and scale drawings
A
1. c
2. a
3. d
4. b

B
1. 10 km
2. a)

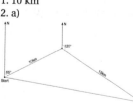

   b) 11.1 cm = 22.2 km
   c) 099°
3. false

C
1. a) i) 325 m
   ii) 060°
   iii) 120°
   b)

---

2. Lengths must be ± 2 mm.

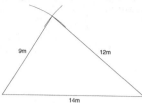

Transformations 1
A
1. a
2. c
3. b
4. b

B
1. a)
   b)
   c)

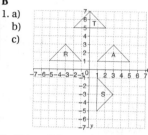

2. a) Translation   b) Rotation
   c) Translation   d) Reflection

C
1. a) Reflection in the x axis.
   b) Rotation 90° anticlockwise about (0, 0).
2.

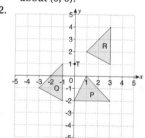

Transformations 2
A
1. d
2. b
3. b

B
1. a) Reflection in the y axis
   b) Rotation 90° clockwise about (0, 0)
   c) Reflection in the line $y = x$
2.

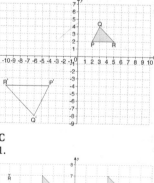

C
1.

---

2. a) Translation by the vector $\begin{pmatrix} -12 \\ -4 \end{pmatrix}$
   b) Enlargement by a scale factor of $-\frac{1}{2}$, centre of enlargement at (0, 5)
   c) Reflection in the line $y = 0$

Similarity and congruency
A
1. c
2. b
3. b
4. b

B
1. a) $n = 7.5$ cm
   b) $n = 6.5$ cm
   c) $n = 6.5$ cm
   d) $n = 7.3$ cm
2. a) Congruent (RHS)
   b) Not congruent
   c) Congruent (SSS)

C
1. a) i) angle XMN = 83°
   ii) angle XNM = XẐY = 68°
   Angles in a triangle add up to 180°. Therefore $180° - 68° - 29° = 83°$
   b) 3.65 cm
   c) 11.74 cm
2. a) 225 cm²
   b) 1.08 litres

Loci and coordinates in 3D
A
1. c
2. d
3. a
4. d
5. b

B
1.

2. R = (3, 0, 3), S = (3, 3, 1), T = (0, 3, 1), U = (0, 1, 1)

C
1.

Angle properties of circles
A
1. c
2. a
3. d

4. a
5. c

B
1. a) $a = 65°$
   b) $a = 18°$
   c) $a = 60°$
   d) $a = 50°$
   e) $a = 82°$
2. John is correct. Angle a is 42° because angles in the same segment are equal.

C
1. a) 66°
   b) The angle at the centre is twice that at the circumference.
   Hence 132° ÷ 2 = 66°
2. a) angle ROQ = 140°
   b) angle PRQ = 70°

Pythagoras' theorem
A
1. d
2. b
3. a
4. c

B
1. a) $n = 15$ cm
   b) $n = 12.6$ cm
   c) $n = 15.1$ cm
   d) $n = 24.6$ cm
2. Since $12^2 + 5^2 = 144 + 25 = 169 = 13^2$, the triangle must be right-angled for Pythagoras' Theorem to be applied.
3. Both statements are true.
   Length of line = $\sqrt{(6^2 + 3^2)} = \sqrt{45}$ in surd form.
   Midpoint = $\dfrac{(2 + 5)}{2}, \dfrac{(11 + 5)}{2}$
   = (3.5, 8)

C
1. $\sqrt{61}$ cm
2. 24.1 cm
3. 13.7 cm
4. $\sqrt{41}$ units

Trigonometry 1
A
1. a
2. b
3. d
4. a
5. c

B
1. a) $n = 5$ cm
   b) $n = 6.3$ cm
   c) $n = 13.8$ cm
   d) $n = 14.9$ cm
   e) $n = 6.7$ cm
2. a) 38.7°
   b) 52.5°
   c) 23.6°

C
1. 15 cm
2. a) 9.9 cm   b) 65°
3. 18 cm

Trigonometry 2
A
1. c
2. a
3. d
4. c
5. b

B
1. 068°
2. 13.8 cm (1 d.p.)
3. a) i) true
   ii) true
   b) 40.9° (1 d.p.)

C
1. 30.1°
2. a) 389.9 m (1d.p.)
   b) 14.6° (1d.p.)
   c) 22.5° (1 d.p.)

Further trigonometry
A
1. c
2. a
3. d
4. c

B
1. a) 13.2 cm (3 s.f.)
   b) 17.3 cm (3 s.f.)
   c) 32.5° (1 d.p.)
   d) 70.9° (1 d.p.)
2. Area = $\frac{1}{2} \times 18 \times 13 \times \sin 37°$
   = 70.4 cm²
   Area is approximately 70 cm². Isobel is correct.
3. a) A (0°, 1), B (90°, 0), C (270°, 0), D (360°, 1)
   b)
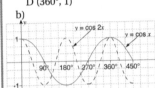

C
1. a) 10.5 cm (3 s.f.)
   b) 12.5 cm² (3 s.f.)
   c) 42.5 cm² (3 s.f.)
2. 4.13 cm

Measures and measurement
A
1. b
2. d
3. a
4. d
5. c

B
1. a) 8000 m
   b) 3.25 kg
   c) 7000 kg
   d) 0.52 m
   e) 2700 ml
   f) 0.002 62 km
2. 12.5 miles
3. 1.32 pounds
4. Lower limit = 46.5 metres
   Upper limit = 47.5 metres
5. 53.3 mph
6. 0.1 g/cm⁻³

C
1. a) 17.6 pounds
   b) 48 kilometres
2. 80 kg
3. a) 1 hour 36 minutes
   b) 4.4 km/h
4. Length = 12.05 cm,
   width = 5.5 cm

Area of 2D shapes
A
1. c
2. d

3. b
4. d
5. b

B
1. a) false
   b) true
   c) true
   d) false
2. 38.6 cm
3. 84.21 cm²
4. 70 000 cm²

C
1. 81 cm²
2. 38.8 cm
3. 16 cm (to nearest cm)
4. 120.24 cm²

Volume of 3D shapes
A
1. a
2. c
3. a
4. c
5. d

B
1. Emily is not correct. The correct volume is 345.6 ÷ 2, i.e. 172.8 cm³.
2. 170.2 cm³
3. 9.9 cm
4. 3807 cm³
5. Volume needs three dimensions.
   $V = 4pr^2$ is only two-dimensional, hence must be a formula for area and not volume.

C
1. 64 cm³
2. a) 672 cm³
   b) 0.000 672 m³
3. 3.2 cm
4. $4r^2p$ volume
   $3\pi\sqrt{(r^2 + p^2)}$ length
   $\dfrac{4\pi r^2}{3p}$ length

Further length, area and volume
A
1. b
2. a
3. d
4. a
5. c

B
1. Solid A is 314 cm³
   Solid B is 600 cm³
   Solid C is 2 145 cm³
   Solid D is 68 cm³
2. The statement is false because:
   Area of segment
   = area of sector − area of triangle
   = 13.09 − 10.83
   = 2.26 cm²

C
1. 73°
2. 0.0248 m³ (3 s.f.)
3. 8 mm

Vectors
A
1. b
2. c
3. a
4. d
5. b

B
1. a) true
   b) false
   c) true
   d) false
2. $\overrightarrow{OC} = 2\mathbf{a} - 3\mathbf{b}$, $\overrightarrow{OD} = 6(2\mathbf{a} - 3\mathbf{b})$
   Hence the vectors are parallel as one vector is a multiple of the other.
   The vectors lie on a straight line through O.
3.

   a) $\mathbf{a} + \mathbf{b} = \begin{pmatrix} -1 \\ 8 \end{pmatrix}$
   b) $\mathbf{a} - \mathbf{b} = \begin{pmatrix} 3 \\ -1 \end{pmatrix}$

C
1. $\overrightarrow{AB} = -3\mathbf{a} + 3\mathbf{b}$,
   $\overrightarrow{CD} = -5\mathbf{a} + 5\mathbf{b}$
   Since $\overrightarrow{AB} = \frac{3}{5}\overrightarrow{CD}$, AB and CD are parallel.
2. a) $\overrightarrow{AC} = \mathbf{a} + \mathbf{b}$
   b) $\overrightarrow{BD} = \mathbf{b} + 2\mathbf{a} - \mathbf{a}$
      = $3\mathbf{b} - \mathbf{a}$
      hence $\overrightarrow{AD} = \overrightarrow{AB} + \overrightarrow{BD}$
         = $\mathbf{a} + 3\mathbf{b} - \mathbf{a}$
      $\overrightarrow{AD} = 3\mathbf{b}$
      $\overrightarrow{AD} = 3\overrightarrow{BC}$, so BC is parallel to AD.
   c) $\overrightarrow{AN} = \overrightarrow{AD} + \overrightarrow{DN}$
      = $3\mathbf{b} - \frac{1}{2}(2\mathbf{b} - \mathbf{a})$
      = $3\mathbf{b} - \mathbf{b} + \frac{1}{2}\mathbf{a}$
      = $\frac{1}{2}\mathbf{a} + 2\mathbf{b}$
   d) $\overrightarrow{YD} = \overrightarrow{YA} + \overrightarrow{AD}$
      = $-\frac{3}{4}(\frac{1}{2}\mathbf{a} + 2\mathbf{b}) + 3\mathbf{b}$
      = $-\frac{3}{8}\mathbf{a} - \frac{3}{2}\mathbf{b} + 3\mathbf{b}$
      = $-\frac{3}{8}\mathbf{a} + \frac{3}{2}\mathbf{b}$
      = $\frac{3}{8}(4\mathbf{b} - \mathbf{a})$

HANDLING DATA
Collecting data
A
1. b
2. d
3. a
4. b
5. d

B
1. The tick boxes overlap. Which box would somebody who did 2 hours of homework tick? It also needs an extra box with 5 or more hours.

   How much time do you spend, to the nearest hour, doing homework each night?

   0 up to 1 hour
   1 up to 2 hours
   2 up to 3 hours
   4 up to 5 hours
   5 hours or more

2. Year   7:   15 students
          8:        22 students
          9:        20 students
         10:        24 students
         11:        19 students

C

1. From the list below, tick your favourite chocolate bar.

Mars ☐
Twix ☐
Toblerone ☐
Galaxy ☐
Bounty ☐
Snickers ☐
other ....

2. The key to this question is to break it into subgroups.

a) On average, how many hours per school day do you watch television?

0 up to 1 hour ☐
1 up to 2 hours ☐
2 up to 3 hours ☐
3 up to 4 hours ☐
Over 4 hours ☐

b) On average, how many hours at the weekend do you watch television?

0 up to 2 hours ☐
2 up to 4 hours ☐
4 up to 6 hours ☐
6 up to 8 hours ☐
Over 8 hours ☐

3. a) 30 students
   b) 14 girls

Scatter diagrams and correlation

A

1. c
2. a
3. b

B

1. a) Positive correlation
   b) Negative correlation
   c) Positive correlation
   d) No correlation

2. a) Positive correlation
   b)
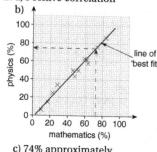
   c) 74% approximately

C

1. a)
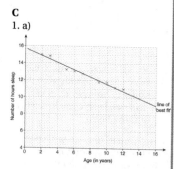
   b) Negative correlation – the younger the child, the more hours sleep they needed.
   c) Line of best fit on diagram above.
   d) A four-year-old child needs approximately 14 hours sleep.
   e) This only gives an estimate as it follows the trend of the data. Similarly, if you continued the line it would assume that you

may eventually need no hours sleep at a certain age, which is not the case.

Averages 1

A

1. c
2. b
3. d
4. d
5. b

B

1. a) false
   b) true
   c) false
   d) true

2. a) mean = 141.35
   b) The manufacturer is justified in making this claim since the mean is just over 141, and the mode and median are also approximately 141.

3. $x = 17$

C

1. 4.65
2. 81
3. £440
4. $3.\dot{3}, 4.\dot{6}, 3.\dot{6}, 3, 2.\dot{3}$

Averages 2

A

1. c
2. b
3. a
4. a

B

1. 21.5 mm
2. a) 47   b) 35   c) 40

C

1.
```
1 | 2 4 9 5 7 5 8 8
2 | 2 7 3 5 7 7
3 | 1 6 5 2 8
4 | 1
```
Reordering gives this.
```
1 | 2 4 5 5 7 8 8 9
2 | 2 3 5 7 7 7
3 | 1 2 5 6 8
4 | 1
```
Key: 1|2 means 12
Stem: 10 minutes

2. a) £31.80
   b) This is only an estimate because the midpoint of the data has been used.
   c) $30 \le x < 40$

Cumulative frequency graphs

A

1. c
2. d
3. c
4. a

B

1. a)

Examination mark	Frequency	Cumulative frequency
0–10	4	4
11–20	6	10
21–30	11	21
31–40	24	45
41–50	18	63
51–60	7	70
61–70	3	73

b)
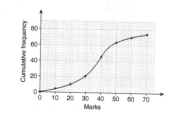

c) 36.5
d) $43 - 28 = 15$ marks
e) 45.5 marks

C

1. a)

Time (nearest minute)	Frequency	Cumulative frequency
$120 < t \le 140$	1	1
$140 < t \le 160$	8	9
$160 < t \le 180$	24	33
$180 < t \le 200$	29	62
$200 < t \le 220$	10	72
$220 < t \le 240$	5	77
$240 < t \le 260$	3	80

b)
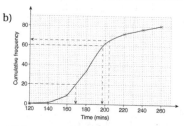

c) i) Interquartile range = $198 - 169 = 29$ minutes
   ii) $80 - 65 = 15$ runners
d)
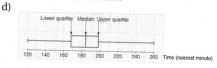

Histograms

A

1. d
2. b
3. b
4. a

B

1. a) $5 < t \le 15$.   26
      $20 < t \le 30$ 14

b)
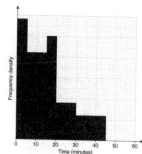

C

1.
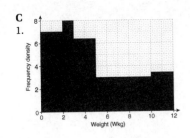

2.

Length S (in seconds)	Frequency
$0 \leq S < 10$	2
$10 \leq S < 15$	5
$15 \leq S < 20$	21
$20 \leq S < 40$	28
$S \geq 40$	0

Probability

A

1. d
2. c
3. b
4. a
5. d

B

1. a)
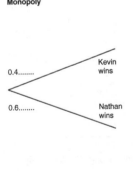

    b) i) $\frac{4}{16} = \frac{1}{4}$
    ii) $\frac{2}{16} = \frac{1}{8}$
    iii) 0

2. 0.09

3. a) $\frac{49}{169}$
    b) $\frac{84}{169}$

C

1. a) i) 0.35
      ii) 0
  b) 50 red beads

2. a) $\frac{6}{36} = \frac{1}{6}$
    b) $\frac{4}{36} = \frac{1}{9}$

3. a)
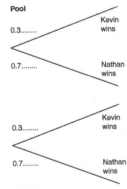

  b) 0.42
  c) 0.46

ACKNOWLEDGEMENTS

The author and publisher are grateful to the copyright
holders for permission to use quoted materials and
photographs.

Letts and Lonsdale
4 Grosvenor Place
London SW1X 7DL

School orders: 015395 64910
School enquiries: 015395 65921
Parent and student enquiries: 015395 64913
Email: enquiries@lettsandlonsdale.co.uk
Website: www.lettsandlonsdale.com

First published 2006
03/130608

ISBN: 9781843156598

British Library Cataloguing in Publication Data. A CIP
record of this book is available from the British Library.

Book concept and development: Helen Jacobs,
Publishing Director

Editorial: Marion Davies and Alan Worth

Author: Fiona C. Mapp

Cover design: Angela English

Inside concept design: Starfish Design

Text design, layout and editorial: Servis Filmsetting

Printed in Italy

Letts and Lonsdale make every effort to ensure that all
paper used in our books is made from wood pulp
obtained from well-managed forests.

C These are GCSE-style questions. Answer all parts of the questions. Show your workings (on separate paper if necessary) and include the correct units in your answers.

1 Enlarge triangle N by scale factor $\frac{1}{3}$ with centre R (−6, 7). (3 marks)

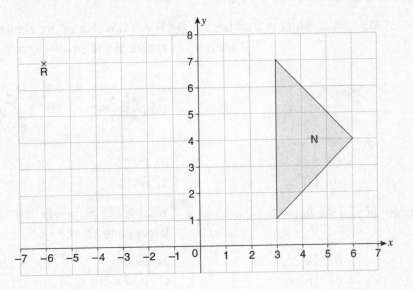

2 The diagram shows four triangles, T_1, T_2, T_3 and T_4.

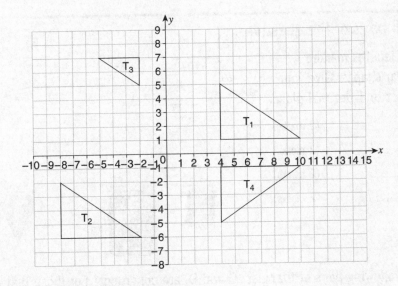

Describe fully the single transformation that maps:

a) T_1 onto T_2 .... (2 marks)

b) T_1 onto T_3 .... (2 marks)

c) T_1 onto T_4 .... (2 marks)

Score / 9

How well did you do? ✗ 1–4 Try again   5–9 Getting there   10–16 Good work   17–21 Excellent! ✓

For more information on this topic, see pages 62–63 of your Success Guide.

49

Similarity & congruency

A Choose just one answer, a, b, c or d.

1 These two shapes are similar. What is the size of angle x? **(1 mark)**

a) 90°   b) 47°
c) 53°   d) 50°

2 What is the length of y in the larger triangle above? **(1 mark)**

a) 14 cm   b) 12 cm
c) 8 cm   d) 16 cm

3 These two shapes are similar. What is the radius of the smaller cone? **(C)** **(1 mark)**

Diagrams not drawn to scale

a) 2 cm   b) 3 cm
c) 4 cm   d) 5 cm

4 What is the perpendicular height of the larger cone above? **(C)** **(1 mark)**

a) 12 cm   b) 9.16 cm
c) 10.47 cm d) 10 cm

Score   / 4

B Answer all parts of the questions.

1 Calculate the lengths marked n in these similar shapes. Give your answers correct to 1 decimal place. **(C)**

a) $n = $ .... **(2 marks)**

b) $n = $ .... **(2 marks)**

c) $n = $ .... **(2 marks)**

d) $n = $ .... **(2 marks)**

a)
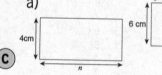
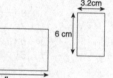

b)
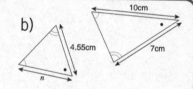

c)
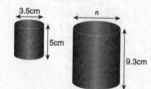

d)
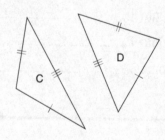

2 Which of the following pairs of triangles, C and D, are congruent? For those that are, state whether the reason is SSS, RHS, SAS or AAS. **(3 marks)**

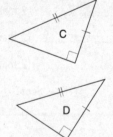
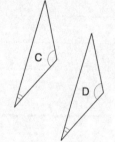

a) ....   b) ....   c) ....

Score   / 11

(C) *Indicates that a calculator may be used*

C

These are GCSE-style questions. Answer all parts of the questions. Show your workings (on separate paper if necessary) and include the correct units in your answers.

1 In the diagram MN is parallel to YZ.

YMX and ZNX are straight lines.

XM = 5.1 cm, XY = 9.5 cm, XN = 6.3 cm, YZ = 6.8 cm

∠YXZ = 29°, ∠XZY = 68°

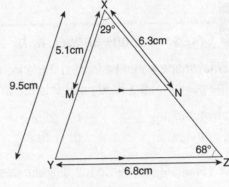

Diagrams not accurately drawn

a) i) Calculate the size of angle XMN. ....° (1 mark)

   ii) Explain how you obtained your answer. ⓒ (1 mark)

   ....

   ....

b) Calculate the length of MN. ⓒ .... (2 marks)

   .... cm

c) Calculate the length of XZ. ⓒ .... (2 marks)

   .... cm

2 Soup is sold in two similar cylindrical cans.

a) The area of the label on the smaller can is 81 cm².

   Calculate the area of the label on the larger can. (The labels are also similar and in the same proportion as the height of the cans.) ⓒ (2 marks)

   ....

b) The capacity of the larger can of soup is 5 litres.

   Calculate the capacity of the smaller can of soup. ⓒ (2 marks)

   ....

Score   / 10

How well did you do? ✗ 1–6 Try again   7–12 Getting there   13–18 Good work   19–25 Excellent! ✓

For more information on this topic, see pages 64–65 of your Success Guide.

51

Loci & coordinates in 3D

A Choose just one answer, a, b, c or d.

1 What shape would be formed if the locus of all the points from a fixed point P is drawn? (1 mark)

a) rectangle   b) square
c) circle   d) kite

Questions 2–5 refer to the diagram opposite.

2 What are the coordinates of point A? (1 mark)

a) (4, 3, 1)   b) (4, 3, 0)
c) (0, 3, 1)   d) (4, 0, 1)

3 What are the coordinates of point B? (1 mark)

a) (0, 3, 1)   b) (0, 0, 0)
c) (4, 3, 0)   d) (0, 3, 0)

4 What are the coordinates of point C? (1 mark)

a) (0, 3, 1)   b) (4, 3, 1)
c) (4, 0, 0)   d) (4, 3, 0)

5 What are the coordinates of point D? (1 mark)

a) (4, 3, 0)   b) (4, 3, 1)
c) (0, 0, 0)   d) (0, 3, 0)

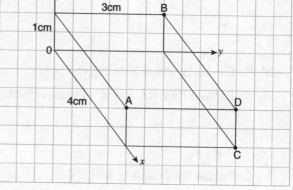

Score / 5

B Answer all parts of the questions.

1 The diagram shows the position of the post office (P), the hospital (H) and the school (S). Robert lives less than 4 miles away from the hospital, less than 5 miles away from the post office and less than 8 miles away from the school.

Show by shading the area where Robert can live. Use a scale of 1 cm = 2 miles. (4 marks)

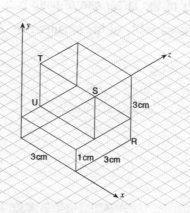

2 The diagram shows a solid. Complete the coordinates for each of the vertices listed below. (4 marks)

R = (.... , ... , ...)
S = (.... , ... , ...)
T = (.... , ... , ...)
U = (.... , ... , ...)

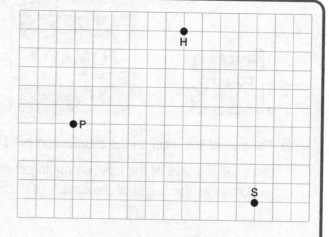

Score / 8

C These are GCSE-style questions. Answer all parts of the questions. Show your workings (on separate paper if necessary) and include the correct units in your answers.

1 In this question you should use ruler and compasses only for the constructions.

Triangle ABC is the plan of an adventure playground, drawn to a scale of 1 cm to 20 m.

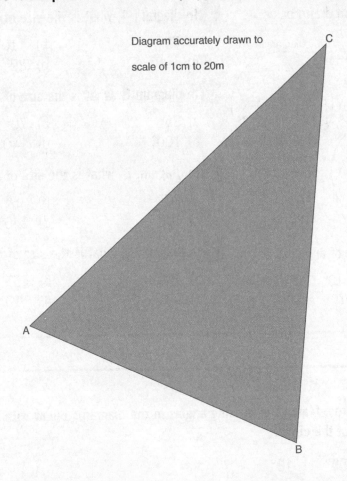

Diagram accurately drawn to

scale of 1cm to 20m

a) On the diagram, draw accurately the locus of the points which are 100 m from C.   (2 marks)

b) On the diagram, draw accurately the locus of the points which are the same distance from A as they are from C.   (2 marks)

c) P is an ice cream kiosk inside the adventure playground.

P is the same distance from A as it is from C.

P is the same distance from AC as it is from AB.

On the diagram, mark the point P clearly with a cross.

Label it with the letter P.   (3 marks)

Score   / 7

How well did you do?   ✗ 1–4 Try again   5–9 Getting there   10–14 Good work   15–20 Excellent! ✓

For more information on this topic, see page 66 of your Success Guide.

53

Angle properties of circles

A

Choose just one answer, a, b, c or d.

Questions 1–5 refer to the diagram drawn below.

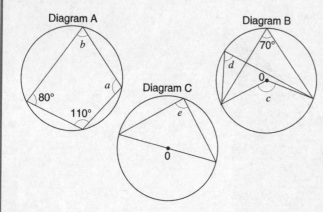

Diagram A

Diagram B

Diagram C

1 In diagram A, what is the size of angle *a*?

a) 80°   b) 110°   (1 mark)

c) 100°   d) 70°

2 In diagram A, what is the size of angle *b*?

a) 70°   b) 110°   (1 mark)

c) 100°   d) 80°

3 In diagram B, what is the size of angle *c*?

a) 35°   b) 70°   (1 mark)

c) 100°   d) 140°

4 In diagram B, what is the size of angle *d*?

a) 70°   b) 35°   (1 mark)

c) 100°   d) 140°

5 In diagram C, what is the size of angle *e*?

a) 100°   b) 45°   (1 mark)

c) 90°   d) 110°

Score   / 5

B

Answer all parts of the questions.

1 Some angles are written on cards. Match the missing angles in the diagrams below with the correct card. O represents the centre of the circle.

(5 marks)

| 50° | 60° | 82° | 65° | 18° |

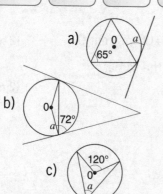

a)

b)

c)

d)

e)

2 John says that 'Angle *a* is 42°.'

Explain whether John is correct.

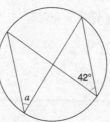

(1 mark)

....

Score   / 6

C These are GCSE-style questions. Answer all parts of the questions. Show your workings (on separate paper if necessary) and include the correct units in your answers.

1

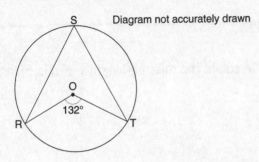

Diagram not accurately drawn

R, S and T are points on the circumference of a circle with centre O.

a) Find angle RST. (1 mark)

.... °

b) Give a reason for your answer. (2 marks)

....

....

....

....

2 PQ and PR are tangents to a circle centre O.

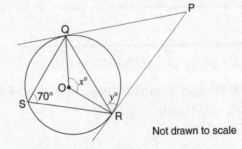

Not drawn to scale

Point S is a point on the circumference.

Angle RSQ is 70°.

a) Find the size of angle ROQ, marked $x°$ in the diagram. (2 marks)

.... °

b) Find the size of angle PRQ, marked $y°$ in the diagram. (4 marks)

.... °

Score   / 9

How well did you do? ✗ 1–2 **Try again**   3–6 **Getting there**   7–12 **Good work**   13–20 **Excellent!** ✓

For more information on this topic, see page 67 of your Success Guide.

Pythagoras' theorem

A Choose just one answer, a, b, c or d.

1 What is the name of the longest side of a right-angled triangle? **(1 mark)**

a) hypo
b) hippopotamus
c) crocodile
d) hypotenuse

2 Calculate the missing length *y* of this triangle.

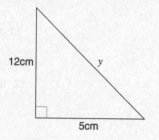

a) 169 cm
b) 13 cm **(1 mark)**
c) 17 cm
d) 84.5 cm

3 Calculate the missing length *y* of this triangle.

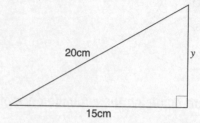

a) 13.2 cm
b) 5 cm **c) (1 mark)**
c) 25 cm
d) 625 cm

4 Point A has coordinates (1, 4), point B has coordinates (4, 10). What are the coordinates of the midpoint of the line AB? **(1 mark)**

a) (5, 14)
b) (3, 6)
c) (2.5, 7)
d) (1.5, 3)

Score   / 4

B Answer all parts of the questions.

1 Calculate the missing lengths of these right-angled triangles. Give your answer to 3 significant figures, where appropriate. **C**

a)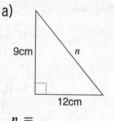

b)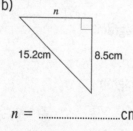

c)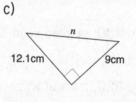

d)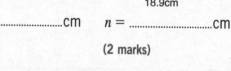

n = ...cm
(2 marks)

n = ...cm
(2 marks)

n = ...cm
(2 marks)

n = ...cm
(2 marks)

2 Molly says, 'The angle *x*° in this triangle is 90°.'

Explain how Molly knows this without measuring the size of the angle.

....

....

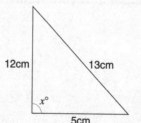

(2 marks)

3 Colin says, 'The length of this line is $\sqrt{45}$ units and the coordinates of the midpoint are (3.5, 8).'

Decide whether these statements are true or false.
Give an explanation for your answer.

....

....

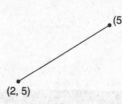

(5, 11)

(2, 5)

(2 marks)

Score   / 12

56

C *Indicates that a calculator may be used*

C These are GCSE-style questions. Answer all parts of the questions. Show your workings (on separate paper if necessary) and include the correct units in your answers.

1 ABC is a right-angled triangle.

AB = 5 cm, BC = 6 cm

Calculate the length of AC.

Leave your answer in surd form.

(3 marks)

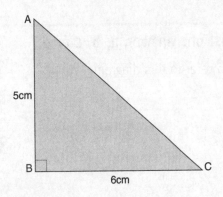

.... cm

2 Calculate the length of the diagonal of this rectangle. Give your answer to one decimal place. (3 marks)

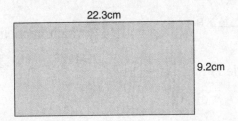

.... cm **C**

3 Calculate the perpendicular height of this isosceles triangle. Give your answer to one decimal place. **C**

(3 marks)

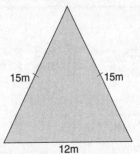

.... m

4 Calculate the length of AB in this diagram. Leave your answer in surd form. (3 marks)

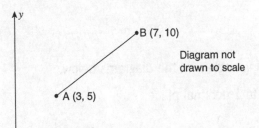

....

Score / 12

How well did you do? ✗ 1–7 Try again   8–14 Getting there   15–21 Good work   22–28 Excellent! ✓

For more information on this topic, see pages **68–69** of your Success Guide.

Trigonometry 1

A Choose just one answer, a, b, c or d.

Questions 1–5 refer to this diagram.

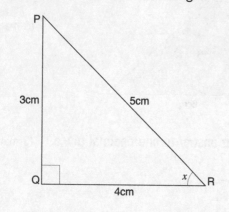

1 Which length is opposite angle *x*? (1 mark)

a) PQ   b) PR   c) QR   d) RX

2 Which length of the triangle is the hypotenuse? (1 mark)

a) PQ   b) PR   c) QR   d) RX

3 Which fraction represents tan *x*? (1 mark)

a) $\frac{3}{5}$ b) $\frac{4}{3}$ c) $\frac{4}{5}$ d) $\frac{3}{4}$

4 Which fraction represents sin *x*? (1 mark)

a) $\frac{3}{5}$ b) $\frac{4}{3}$ c) $\frac{5}{3}$ d) $\frac{4}{5}$

5 Which fraction represents cos *x*? (1 mark)

a) $\frac{3}{5}$ b) $\frac{3}{4}$ c) $\frac{4}{5}$ d) $\frac{5}{4}$

Score   / 5

B Answer all parts of the questions.

1 Choose a card for each of the missing lengths *n* on the triangles. The lengths have been rounded to 1 decimal place. **C**

| 6.3 cm | 6.7 cm | 13.8 cm | 5 cm | 14.9 cm |

(5 marks)

      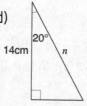 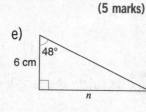

a)   b)   c)   d)   e)

n = ... cm   *n* = ... cm   *n* = ... cm   *n* = ... cm   *n* = ... cm

2 Work out the missing angle *x* in the diagrams below.

Give your answers to 1 decimal place. **C**

a)   b)   c)

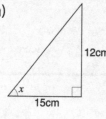 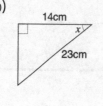   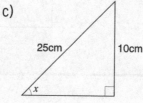

x = ...°   *x* = ...°   *x* = ...°

(2 marks)   (2 marks)   (2 marks)

Score   / 11

C *Indicates that a calculator may be used*

These are GCSE-style questions. Answer all parts of the questions. Show your workings (on separate paper if necessary) and include the correct units in your answers.

1 RS and SU are two sides of a rectangle.

T is a point on SU.

SU is 50 cm.

ST is 18 cm.

Angle STR is 40°.

Calculate the width of the rectangle. **C**

Give your answer correct to the nearest centimetre.

.... cm

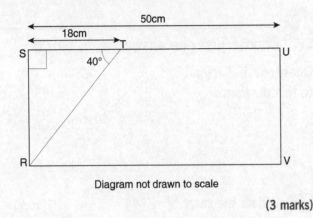

Diagram not drawn to scale

(3 marks)

2 The diagram shows two triangles, PQR and QRS.

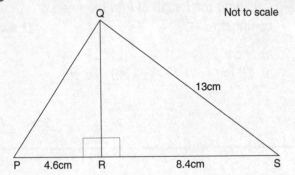

Not to scale

a) Calculate the length of QR. **C** (2 marks)

.... cm

b) Calculate angle QPR. **C** (3 marks)

.... °

3 The diagram shows a right-angled triangle ABC.

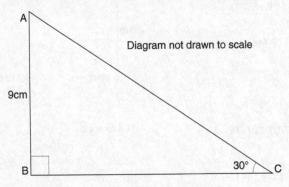

Diagram not drawn to scale

Calculate the length of AC. **C**

.... cm

(3 marks)

Score / 11

How well did you do? ✗ 1–7 Try again   8–12 Getting there   13–19 Good work   20–27 Excellent! ✓

For more information on this topic, see pages 70–71 of your Success Guide.

59

TRIGONOMETRY 1 Shape, Space and Measures

Trigonometry 2

A

Choose just one answer, a, b, c or d.

Questions 1–3 refer to this diagram.

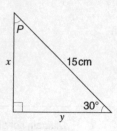

Questions 4 and 5 refer to this diagram.

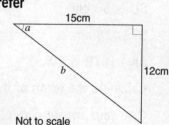

Not to scale

1 Calculate the value of x. (c) (1 mark)

   a) 13 cm   b) 8.7 cm
   c) 7.5 cm   d) 10 cm

2 Calculate the value of y. (c) (1 mark)

   a) 13 cm   b) 8.7 cm
   c) 7.5 cm   d) 10 cm

3 Which fraction represents cos P? (c) (1 mark)

   a) $\frac{13}{15}$ b) $\frac{3}{4}$
   c) $\frac{3}{5}$ d) $\frac{1}{2}$

4 Calculate the size of angle a to the nearest degree. (c) (1 mark)

   a) 53°   b) 45°   c) 39°   d) 37°

5 Calculate the length of b to the nearest centimetre. (c) (1 mark)

   a) 9 cm   b) 19 cm
   c) 29 cm   d) 16 cm

Score /5

B

Answer all parts of the questions.

1 A ship sails 20 km due north and then 50 km due east. What is the bearing of the finishing point from the starting point? (c) (2 marks)

Bearing = ...°

2 The diagram represents the sector of a circle with centre O and radius 12 cm. Angle POR equals 70°.

Calculate the length of the straight line PR. (c) (3 marks)

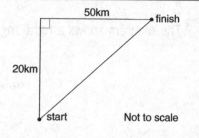

Not to scale

.... cm

3 CDEFGH is a right-angled triangular prism.

N is the midpoint of DE.

a) Decide whether the following statements are true or false.

   i) The length HD is 19.2 cm, correct to 3 s.f. ....

   ii) The size of angle HDC is 51.3°, correct to 1 decimal place. .... (2 marks)

b) Calculate the size of angle HNC. ....° (c) (3 marks)

Score / 10

(c) *Indicates that a calculator may be used*

These are GCSE-style questions. Answer all parts of the questions. Show your workings (on separate paper if necessary) and include the correct units in your answers.

1 The diagram shows a triangle PQR.

PS = 6.5 cm, QR = 12.7 cm and angle QRS = 62°.

Calculate the size of the angle marked $x°$.

Give your answer correct to 1 decimal place. **C**

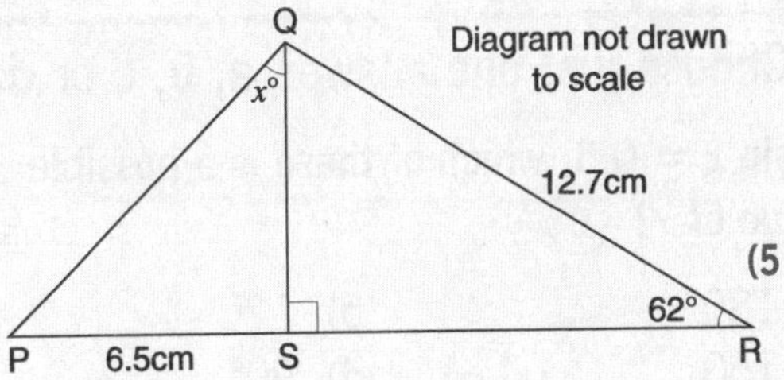

Diagram not drawn to scale

12.7cm

62°

Q $x°$

P   6.5cm   S   R

(5 marks)

....

.... °

2 The angle of elevation of P from T is 18°.

R is 1500 m due west of S and
T is 1200 m due south of S.
SP is a vertical tower.

R, S and T are three points on horizontal ground.

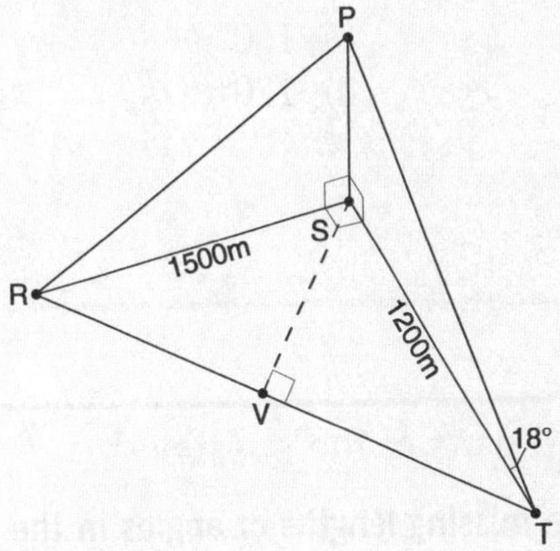

P

1500m

1200m

R   S   V   18°   T

a) Calculate the height of the tower. **C**

(2 marks)

....

.... m

b) Find the angle of elevation of P from R.

(2 marks)

....

.... °

c) V is a point on RT which is nearest to S.

Calculate the angle of elevation of P from V.

(5 marks)

....

....

....

.... °

Score   / 14

How well did you do?   ✗ 1–8 Try again   9–14 Getting there   15–21 Good work   22–29 Excellent! ✓

For more information on this topic, see pages 70–73 of your Success Guide.

Further trigonometry

A

Choose just one answer, a, b, c or d.

1 If sin x = 0.5, which of these is a possible value of x? **c** (1 mark)

a) 180°   b) 120°
c) 150°   d) 90°

2 If cos x = 0.5, which of these is a possible value of x? **c** (1 mark)

a) 300°   b) 150°
c) 65°   d) 120°

3 If sin $x = \frac{\sqrt{3}}{2}$, which of these is a possible value of x? **c** (1 mark)

a) 320°   b) 400°
c) 360°   d) 420°

4 Which of these is the correct formula for the cosine rule? (1 mark)

a) $a^2 = b^2 - c^2 + 2bc \cos A$
b) $b^2 = a^2 + c^2 - 2bc \cos B$
c) $a^2 = b^2 + c^2 - 2bc \cos A$
d) $c^2 = b^2 + a^2 - 2ab \cos A$

Score   / 4

B

Answer all parts of the questions.

1 Calculate the missing lengths or angles in the diagrams below. **c**

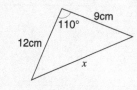

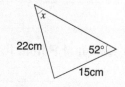

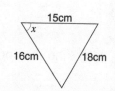

a) x = ... cm   b) x = ... cm   c) x = ...°   d) x = ...°
   (2 marks)   (2 marks)   (2 marks)   (2 marks)

2 Isobel says, 'The area of this triangle is 70 cm².'

Decide, with working to justify your answer, **c** whether this statement is true or false.

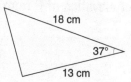

(2 marks)

....

....

3 The diagram shows a sketch of part of the curve $y = f(x)$, where $f(x) = \cos x°$.

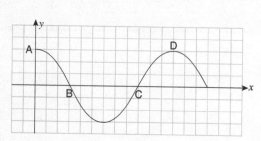

a) Write down the coordinates of the points:

   A (........, ...)   B (........, ...)   C (........, ...)   D (........, ...)   (4 marks)

b) On the same diagram, sketch the graph of $y = \cos 2x$ (3 marks)

Score   / 17

c Indicates that a calculator may be used

C These are GCSE-style questions. Answer all parts of the questions. Show your workings (on separate paper if necessary) and include the correct units in your answers.

1 The diagram shows a quadrilateral PQRS.

PS = 4.3 cm, PQ = 5.8 cm, SR = 7.3 cm

Angle PSR = 127°, angle PQR = 54°

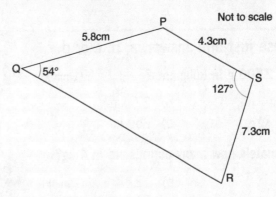

Not to scale

a) Calculate the length of PR. Ⓒ

Give your answer correct to 3 significant figures. .... (3 marks)

.... cm

b) Calculate the area of triangle PRS. Ⓒ

Give your answer correct to 3 significant figures. .... (2 marks)

.... cm²

c) Calculate the area of the quadrilateral PQRS. Ⓒ

Give your answer correct to 3 significant figures. .... (6 marks)

....

.... cm²

2

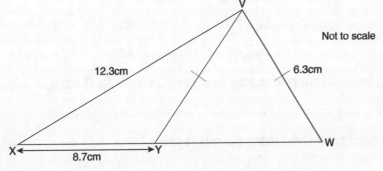

Not to scale

In triangle VWX, Y is the point on WX such that VW = VY.

VW = 6.3 cm, VX = 12.3 cm, XY = 8.7 cm

Calculate the length of WY. Ⓒ   (5 marks)

....

.... cm

Score   / 16

How well did you do?   ✗ 1–10 Try again   11–19 Getting there   20–27 Good work   28–37 Excellent! ✓

For more information on this topic, see pages 74–75 & 78 of your Success Guide.

63

Measures & measurement

A Choose just one answer, a, b, c or d.

1 What is 2 500 g in kilograms? (1 mark)

- a) 25 kg
- b) 2.5 kg
- c) 0.25 kg
- d) 250 kg

2 Approximately how many pounds are in 4 kg?

- a) 6.9
- b) 12.4 (1 mark)
- c) 7.7
- d) 8.8

3 Jessica is 165 cm tall to the nearest cm. What is the lower limit of her height? (1 mark)

- a) 164.5 cm
- b) 165.5 cm
- c) 165 cm
- d) 164.9 cm

4 What is the volume of a piece of wood with density 680 kg m⁻³ and mass 34 kg? (1 mark)

- a) 0.5 m³
- b) 20 m³ (c)
- c) 2 m³
- d) 0.05 m³

5 A car travels for two and a half hours at a speed of 42 mph. How far does the car travel? (c) (1 mark)

- a) 96 miles
- b) 100 miles
- c) 105 miles
- d) 140 miles

Score   / 5

B Answer all parts of the questions.

1 Complete the statements below. (6 marks)

a) 8 km = ... m

b) 3 250 g = ... kg

c) 7 tonnes = ... kg

d) 52 cm = ... m

e) 2.7 litres = ... ml

e) 262 cm = ... km

2 Two towns are approximately 20 km apart. Approximately how many miles is this? (c) (1 mark)

....

3 A recipe uses 600 g of flour. Approximately how many pounds is this? (c) (1 mark)

....

4 A field is 47 metres long to the nearest metre. Write down the upper and lower limits of the length of the field. (2 marks)

....

5 Giovanni drove 200 miles in 3 hours and 45 minutes. At what average speed did he travel? (2 marks)

.... (c)

6 What is the density of a toy if its mass is 200 g and its volume is 2000 cm³? (2 marks)

....

Score   / 14

(c) *Indicates that a calculator may be used*

C These are GCSE-style questions. Answer all parts of the questions. Show your workings (on separate paper if necessary) and include the correct units in your answers.

1 a) Change 8 kilograms into pounds. (2 marks)

.... pounds Ⓒ

b) Change 30 miles into kilometres. (2 marks)

.... km

2 Two solids each have a volume of 2.5m³.

The density of solid A is 320 kg per m³.

The density of solid B is 288 kg per m³.

Calculate the difference in the masses of the solids. Ⓒ .... kg (3 marks)

3 Amy took part in a sponsored walk.

She walked from the school to the park and back.

The distance from the school to the park is 8 km.

a) Amy walked from the school to the park at an average speed of 5 km/h.

Find the time she took to walk from the school to the park. Ⓒ (2 marks)

....

b) Her average speed for the return journey was 4 km/h.

Calculate her average speed for the whole journey. Ⓒ (4 marks)

.... km/h

4 The length of the rectangle is 12.1 cm to the nearest mm.

The width of the rectangle is 6 cm to the nearest cm.

Write down the lower limits for the length and width of the rectangle.

12.1cm

6cm

Length .... cm (2 marks)

Width .... cm

Score / 15

How well did you do?   ✗ 1–11 Try again   12–19 Getting there   20–27 Good work   28–34 Excellent! ✓

For more information on this topic, see pages 76–77 of your Success Guide.

65

Area of 2D shapes

A Choose just one answer, a, b, c or d.

1 What is the area of this triangle? (1 mark)

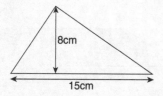

8cm
15cm

a) 60 mm² b) 120 cm²
c) 60 cm² d) 46 cm²

2 What is the approximate circumference of a circle of radius 4 cm? (1 mark)

a) 25.1 cm² b) 50.3 cm
c) 12.6 cm d) 25.1 cm

3 Change 50 000 cm² into m². (1 mark)

a) 500 m² b) 5 m²
c) 0.5 m² d) 50 m²

4 What is the area of this circle? (C) (1 mark)

4cm

a) 25.1 cm² b) 55 cm²
c) 12.6 cm² d) 50.3 cm²

5 If the area of a rectangle is 20 cm² and its width is 2.5 cm, what is the length of the rectangle? (1 mark)

a) 9 cm b) 8 cm
c) 7.5 cm d) 2.5 cm

Score / 5

B Answer all parts of the questions.

1 For each of the diagrams below, decide whether the area given is true or false.

a)
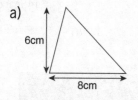
6cm
8cm

b)
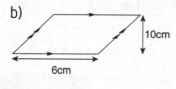
10cm
6cm

c)
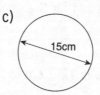
15cm

d)
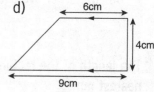
6cm
4cm
9cm

a) Area = 48 cm² .... (1 mark)

b) Area = 60 cm² .... (1 mark)

c) Area = 177 cm² .... (1 mark)

d) Area = 108 cm² .... (1 mark)

2 Calculate the perimeter of this shape. (C) (3 marks)

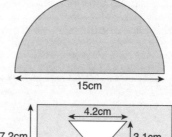
15cm

.... cm

3 Calculate the area of the shaded region. (C) (3 marks)

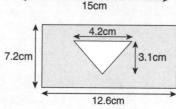
4.2cm
7.2cm
3.1cm
12.6cm

.... cm²

4 Change 7m² to cm². .... cm² (2 marks)

Score / 12

(C) *Indicates that a calculator may be used*

C

These are GCSE-style questions. Answer all parts of the questions. Show your workings (on separate paper if necessary) and include the correct units in your answers.

1 Work out the area of the shape shown in the diagram. (C)

State the units with your answer.

....

....

....

....

....

....

15cm

(5 marks)

8cm

5cm

9cm

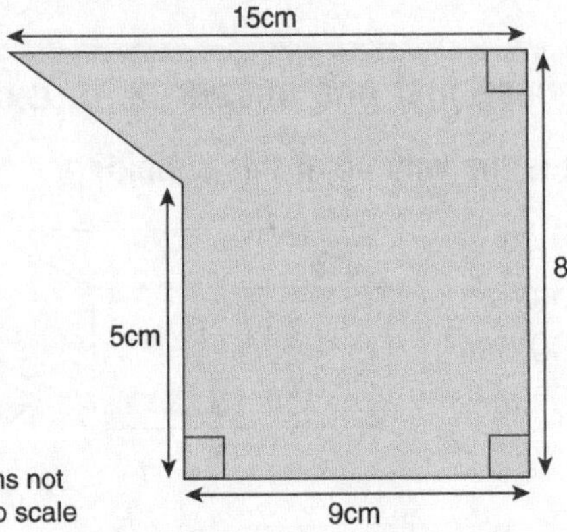

Diagrams not drawn to scale

2 Calculate the perimeter of the shape shown in the diagram. (C)

Give your answer to 3 significant figures.

....

....

....

....

....

.... cm

6cm

10cm

(3 marks)

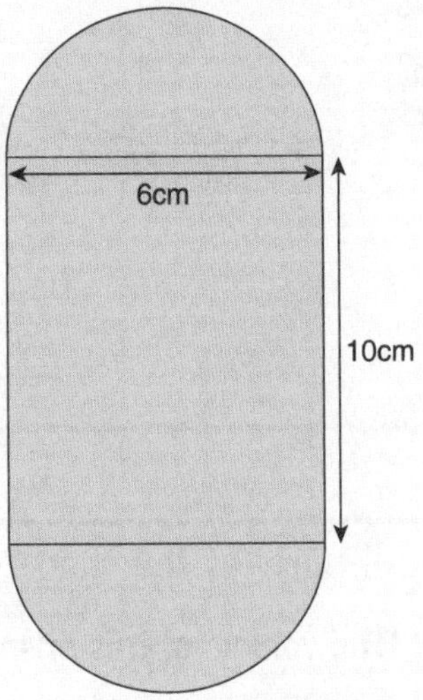

3 The area of a circular sewing pattern is 200 cm².

Calculate the diameter of the sewing pattern.
Give your answer correct to the nearest centimetre. (C)

(4 marks)

....

.... cm

4 Calculate the area of the shape.

State the units with your answer. (C)

....

....

....

....

13.8cm

(3 marks)

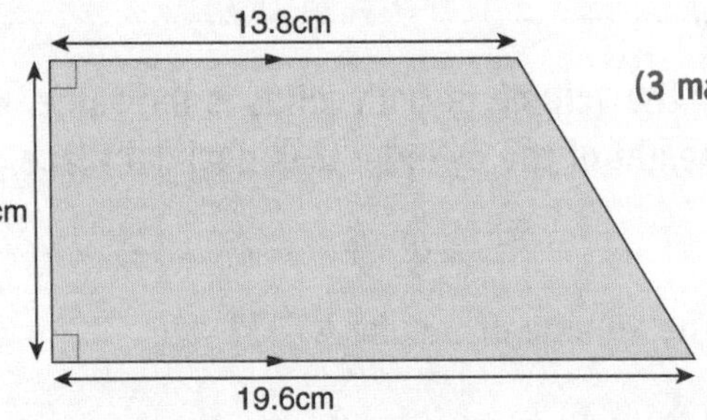

7.2cm

19.6cm

How well did you do?   ✗ 0–8 Try again   9–16 Getting there   17–24 Good work   25–32 Excellent! ✓

For more information on this topic, see pages 78–79 of your Success Guide.

AREA OF 2D SHAPES Shape, Space and Measures

Volume of 3D shapes

A Choose just one answer, a, b, c or d.

1 What is the volume of this cuboid?   (1 mark)

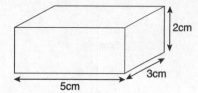

a) 30 cm³
b) 16 cm³
c) 300 mm³
d) 15 cm³

2 What is the volume of this prism?   (1 mark)

a) 64 cm³
b) 240 cm³
c) 120 cm³
d) 20 cm³

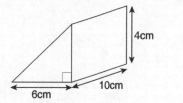

3 The volume of a cuboid is 20 cm³. If its height is 1 cm and its width is 4 cm, what is its length?   (1 mark)

a) 5 cm   b) 10 cm
c) 15 cm   d) 8 cm

4 A cube of volume 2 cm³ is enlarged by a scale factor of 3. What is the volume of the enlarged cube?   (1 mark)

a) 6 cm³   b) 27 cm³
c) 54 cm³   d) 18 cm³

5 If p and q represent lengths, decide what the formula $\frac{3}{5}\pi p^2 q$ shows.   (1 mark)

a) circumference   b) length
c) area   d) volume   Score   / 5

B Answer all parts of the questions.

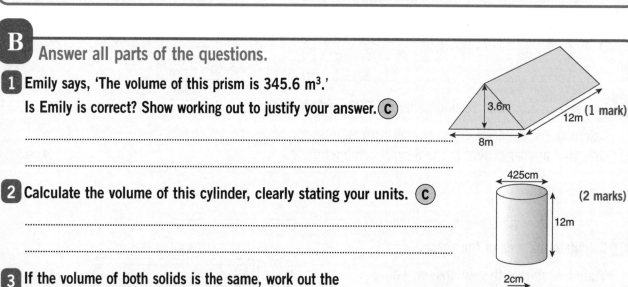

1 Emily says, 'The volume of this prism is 345.6 m³.'

Is Emily is correct? Show working out to justify your answer. Ⓒ   (1 mark)

....

....

2 Calculate the volume of this cylinder, clearly stating your units. Ⓒ   (2 marks)

....

....

3 If the volume of both solids is the same, work out the height of the cylinder to 1 decimal place. Ⓒ   (4 marks)

....**cm**

4 The volume of a cube is 141 cm³.

Each length of the cube is enlarged by a scale factor of 3.
What is the volume of the enlarged cube? Ⓒ   (2 marks)

.... **cm³**

5 Lucy says that 'The volume of a sphere is given by the formula $V = 4\pi r^2$'.
Explain why she cannot be correct. Ⓒ   (1 mark)

....   Score   / 10

Ⓒ *Indicates that a calculator may be used*

C These are GCSE-style questions. Answer all parts of the questions. Show your workings (on separate paper if necessary) and include the correct units in your answers.

1 A cube has a surface area of 96 cm². Work out the volume of the cube. (4 marks)

....

2 A door wedge is in the shape of a prism with cross section **VWXY**.

VW = 7 cm, VY = 15 cm, WX = 9 cm.

The width of the door wedge is 0.08 m.

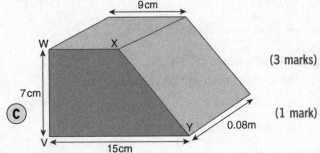

a) Calculate the volume of the door wedge. (C) (3 marks)

.... cm³

b) What is the volume of the door wedge in m³? (C) (1 mark)

.... m³

3 The volume of this cylinder is 250 cm³.

The height of the cylinder is 8 cm.

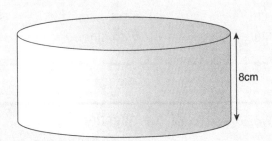

Calculate the radius of the cylinder giving your answer to 1 decimal place. (C) (3 marks)

.... cm

4 Here are three expressions.

Expression	Length	Area	Volume
$4r^2p$			
$3\pi\sqrt{(r^2+p^2)}$			
$\dfrac{4\pi r^2}{3p}$			

r and p are lengths.

Put a tick in the correct column to show whether the expression can be used for length, area or volume. (3 marks)

Score   / 14

How well did you do? ✗ 1–7 Try again   8–14 Getting there   15–22 Good work   23–29 Excellent! ✓

For more information on this topic, see pages 80–81 & 62 of your Success Guide.

69

Further length, area & volume

A Choose just one answer, a, b, c or d.

1 A sphere has a radius of 3 cm. What is the volume of the sphere given in terms of π? *(1 mark)*

a) 12π b) 36π c) 42π d) $\frac{81}{4}\pi$

2 A sphere has a radius of 4 cm. What is the surface area of the sphere, given in terms of π? *(1 mark)*

a) 64π b) 32π c) $\frac{256}{3}\pi$ d) 25π

3 The volume of a pyramid is 25 cm³. The area of the base is 12 cm². What is the perpendicular height of the pyramid? *(1 mark)*

a) 7.2 cm   b) 4 cm

c) 25 cm   d) 6.25 cm

Questions 4 and 5 refer to this circle diagram

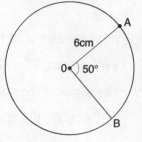

4 What is the length of arc AOB? Ⓒ *(1 mark)*

a) 5.2 cm   b) 5.8 cm
c) 6.2 cm   d) 7.4 cm

5 What is the area of sector AOB? Ⓒ *(1 mark)*

a) 15.2 cm²   b) 25.3 cm²
c) 15.7 cm²   d) 16.9 cm²

Score   / 5

B Answer all parts of the questions.

1 The volumes of the solids below have been calculated. Match each solid with its correct volume. Ⓒ

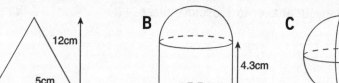

A

B

C

D

600 cm³.........................   68 cm³.........................   314 cm³.........................   2 145 cm³.........................
(2 marks)   *(2 marks)*   *(2 marks)*   *(2 marks)*

2 Decide whether this statement is true or false. Ⓒ

You must show sufficient working in order to justify your answer.

'The area of the shaded segment is 3.26 cm².'

(3 marks)

....

....

....

Score   / 11

Ⓒ *Indicates that a calculator may be used*

C These are GCSE-style questions. Answer all parts of the questions. Show your workings (on separate paper if necessary) and include the correct units in your answers.

1 The sector area of a circle is 51.6 cm².

The radius of the circle is 9 cm.

Work out the size of the angle θ of the sector. Ⓒ

Give your answer to the nearest degree.

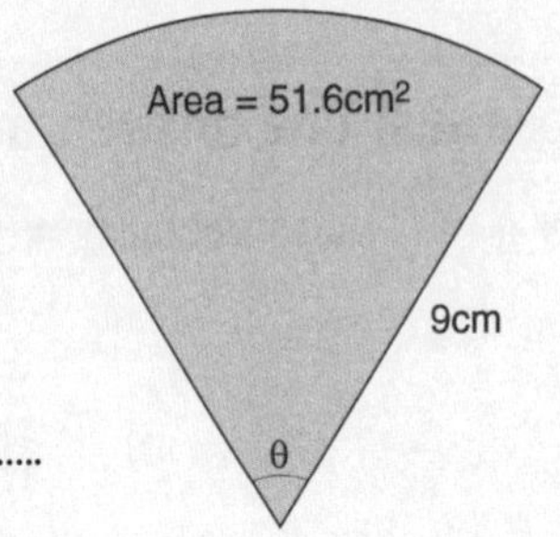

Area = 51.6cm²

9cm

θ

(3 marks)

....

....

....

.... °

2 The diagram shows a plastic container.

The container is formed by joining a cylindrical tube to a hemisphere.

The diameter of the cylinder and hemisphere is 26 cm.

The total height of the container is 51 cm.

Work out the volume of the container. Ⓒ

Give your answer correct to 3 significant figures.

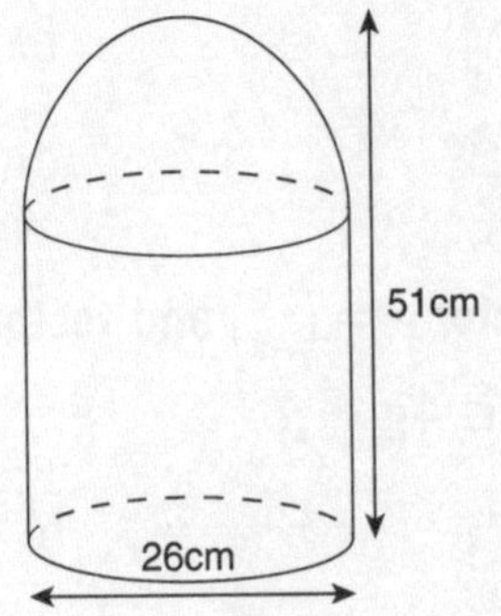

51cm

26cm

(4 marks)

....

....

....

.... m³

3 The volume of a ball bearing is 268 mm³.

Work out the diameter of the ball bearing, giving your answer to the nearest whole number. Ⓒ

(3 marks)

....

....

....

.... mm

Score   / 10

How well did you do?   ✗ 1–7 Try again   8-12 Getting there   3–19 Good work   20–26 Excellent! ✓

For more information on this topic, see pages 82–83 of your Success Guide.

71

Vectors

A Choose just one answer, a, b, c or d.

1 If vector $\underline{a} = \begin{pmatrix} 2 \\ 3 \end{pmatrix}$ and vector $\underline{b} = \begin{pmatrix} -5 \\ -2 \end{pmatrix}$, what is $\underline{a} + \underline{b}$? **(1 mark)**

a) $\begin{pmatrix} 1 \\ -3 \end{pmatrix}$ b) $\begin{pmatrix} -3 \\ 1 \end{pmatrix}$

c) $\begin{pmatrix} 7 \\ -5 \end{pmatrix}$ d) $\begin{pmatrix} -10 \\ -6 \end{pmatrix}$

2 If vector $\underline{c} = \begin{pmatrix} -4 \\ 2 \end{pmatrix}$ and vector $\underline{d} = \begin{pmatrix} -6 \\ -5 \end{pmatrix}$, what is $\underline{c} - \underline{d}$? **(1 mark)**

a) $\begin{pmatrix} 8 \\ -14 \end{pmatrix}$ b) $\begin{pmatrix} -6 \\ -5 \end{pmatrix}$

c) $\begin{pmatrix} 2 \\ 7 \end{pmatrix}$ d) $\begin{pmatrix} -6 \\ 5 \end{pmatrix}$

3 If vector $\underline{p} = \begin{pmatrix} 7 \\ -2 \end{pmatrix}$ and vector $\underline{r} = \begin{pmatrix} -9 \\ 2 \end{pmatrix}$, what is $4\underline{p} + \underline{r}$? **(1 mark)**

a) $\begin{pmatrix} 19 \\ -6 \end{pmatrix}$ b) $\begin{pmatrix} 38 \\ 0 \end{pmatrix}$

c) $\begin{pmatrix} -6 \\ 19 \end{pmatrix}$ d) $\begin{pmatrix} 20 \\ -3 \end{pmatrix}$

4 If vector $\underline{t} = \begin{pmatrix} 3 \\ -2 \end{pmatrix}$, what is the magnitude of $\underline{t}$? **(1 mark)**

a) $\sqrt{5}$ b) $\sqrt{10}$

c) 13 d) $\sqrt{13}$

5 Which vector is parallel to vector $\underline{r} = \begin{pmatrix} -4 \\ 6 \end{pmatrix}$? **(1 mark)**

a) $\begin{pmatrix} -8 \\ 6 \end{pmatrix}$ b) $\begin{pmatrix} -16 \\ 24 \end{pmatrix}$

c) $\begin{pmatrix} -4 \\ 12 \end{pmatrix}$ d) $\begin{pmatrix} -8 \\ 18 \end{pmatrix}$

Score   / 5

B Answer all parts of the questions.

1 The statements below refer to the diagram opposite.

Decide whether the statements are true or false. **(4 marks)**

a) $\overrightarrow{OB} = a + b$ ....

b) $\overrightarrow{BC} = -a + b + c$ ....

c) $\overrightarrow{AC} = -a + c$ ....

d) If N is the midpoint of OC then $\overrightarrow{AN} = -\frac{1}{2}c + a$ ....

2 If $\overrightarrow{OC} = 2\underline{a} - 3\underline{b}$ and $\overrightarrow{OD} = 12\underline{a} - 18\underline{b}$, write down two geometrical facts about the vectors $\overrightarrow{OC}$ and $\overrightarrow{OD}$. **(2 marks)**

....

....

3 On grid paper, draw the vectors $\underline{a}$ and $\underline{b}$, then complete the statements. **(4 marks)**

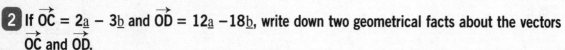

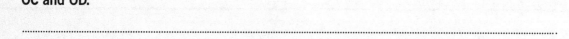

a) $a = \begin{pmatrix} 3 \\ 2 \end{pmatrix}$, $b = \begin{pmatrix} -4 \\ 6 \end{pmatrix}$, $a + b = \begin{pmatrix} \\ \end{pmatrix}$

b) $a = \begin{pmatrix} 1 \\ 4 \end{pmatrix}$, $b = \begin{pmatrix} -2 \\ 5 \end{pmatrix}$, $a - b = \begin{pmatrix} \\ \end{pmatrix}$

Score   / 10

These are GCSE-style questions. Answer all parts of the questions. Show your workings (on separate paper if necessary) and include the correct units in your answers.

1

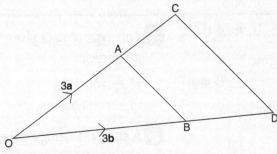

$\overrightarrow{OA} = 3\mathbf{a}$, $\overrightarrow{OB} = 3\mathbf{b}$, $\overrightarrow{OC} = 5\mathbf{a}$, $\overrightarrow{BD} = 2\mathbf{b}$

Prove that AB is parallel to CD. (3 marks)

....

....

2

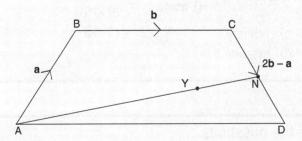

ABCD is a quadrilateral with $\overrightarrow{AB} = \mathbf{a}$, $\overrightarrow{BC} = \mathbf{b}$, and $\overrightarrow{CD} = 2\mathbf{b} - \mathbf{a}$.

a) Express $\overrightarrow{AC}$ in terms of $\mathbf{a}$ and $\mathbf{b}$. .... (1 marks)

b) Prove that BC is parallel to AD. .... (2 marks)

c) N is the midpoint of CD.

Express $\overrightarrow{AN}$ in terms of $\mathbf{a}$ and $\mathbf{b}$. (2 marks)

....

d) Y is the point on AN such that AY : YN = 3 : 1.

Show that $\overrightarrow{YD} = \frac{3}{8}(4\mathbf{b} - \mathbf{a})$.

....

.... (3 marks)

Score   / 11

How well did you do?   ✗ 1–4 **Try again**   5–9 **Getting there**   10–17 **Good work**   18–26 **Excellent!** ✓

Collecting data

A Choose just one answer, a, b, c or d.

1 What is the name given to data you collect yourself? *(1 mark)*

a) continuous   b) primary
c) secondary   d) discrete

2 Data which is usually obtained by counting is said to be this type of data. *(1 mark)*

a) continuous   b) primary
c) secondary   d) discrete

3 This type of data changes from one category to the next. *(1 mark)*

a) continuous   b) primary
c) secondary   d) discrete

4 This type of data gives a word as an answer. *(1 mark)*

a) quantitative   b) qualitative
c) continuous   d) discrete

5 A survey is being carried out on the number of hours some students spend watching television. In year 7 there were 240 students, year 8 had 300 and year 9 460 students. John decides to use a stratified sample of 100 students. How many students should he ask from year 7? *(1 mark)*

a) 46   b) 30
c) 48   d) 24

Score   / 5

B Answer all parts of the questions.

1 Jim and Annabelle are designing a survey to use in the school. One of their questions is shown below.

How much time do you spend doing homework per night?

0–1 hrs	1–2 hrs	2–3 hrs	3–4 hrs

What is the problem with this question? Rewrite the question so that it is improved. *(2 marks)*

....

....

....

2 Laura conducts a survey of the students in her school. She decides to interview 100 students.

Calculate the number of students she should choose from each year group to provide a representative sample. Complete the table below. **C** *(3 marks)*

Year group	Number of students	Number of students in sample
7	120	
8	176	
9	160	
10	190	
11	154	

Score   / 5

C *Indicates that a calculator may be used*

C These are GCSE-style questions. Answer all parts of the questions. Show your workings (on separate paper if necessary) and include the correct units in your answers.

1 Mrs Robinson is going to sell chocolate bars at the school tuck shop. She wants to know what type of chocolate bars pupils like. Design a suitable questionnaire she could use.

(2 marks)

2 Robert is conducting a survey into television habits. One of the questions in his survey is: 'Do you watch a lot of television?'.
His friend Jessica tells him that it is not a very good question.

Write down two ways in which Robert could improve the question. (2 marks)

....

....

....

....

3 The table shows the gender and number of students in each year group.

Year group	Number of boys	Number of girls	Total
7	160	120	280
8	108	132	240
9	158	117	275
10	85	70	155
11	140	110	250

Mark is carrying out a survey about how much pocket money students are given.

He decides to take a stratified sample of 150 students from the whole school.

Calculate how many in the stratified sample should be: (4 marks)

a) students from Year 8 .... students

b) girls from Year 11 .... girls

Score   / 8

How well did you do? ✗ 1–4 Try again   5–9 Getting there   10–13 Good work   14–18 Excellent! ✓

For more information on this topic, see pages 88–89 of your Success Guide.

75

Scatter diagrams and correlation

A

Choose just one answer, a, b, c or d.

1 A scatter diagram is drawn to show the height and weight of some students. What type of correlation is shown? (1 mark)

a) zero   b) negative

c) positive   d) scattered

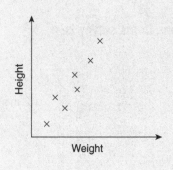

2 A scatter diagram is drawn to show the maths scores and heights of a group of students. What type of correlation is shown? (1 mark)

a) zero

b) negative

c) positive

d) scattered

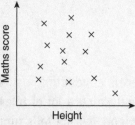

3 A scatter diagram is drawn to show the age of some cars and their values. What type of correlation would be shown? (1 mark)

a) zero   b) negative

c) positive   d) scattered

Score   / 3

B

Answer all parts of the questions.

1 Some statements have been written on cards:

(Positive Correlation)   (Negative Correlation)   (No Correlation)

Decide which card best describes these relationships.

a) The temperature and the sales of ice lollies   .... (1 mark)

b) The temperature and the sales of woollen gloves   .... (1 mark)

c) The weight of a person and his/her waist measurement   .... (1 mark)

d) The height of a person and his/her IQ   .... (1 mark)

2 The scatter diagram shows the marks scored in a Mathematics and Physics examination.

a) Describe the relationship between the Mathematics and Physics scores.

....

b) Draw a line of best fit on the scatter diagram. (1 mark)

c) Use your line of best fit to estimate the Mathematics score that Jonathan is likely to obtain if he has a Physics score of 75%.

....

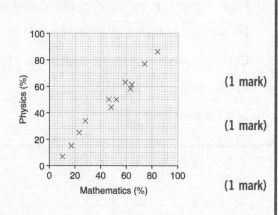

(1 mark)

(1 mark)

(1 mark)

Score   / 7

C These are GCSE-style questions. Answer all parts of the questions. Show your workings (on separate paper if necessary) and include the correct units in your answers.

1 The table shows the ages of some children and the total number of hours sleep they had between noon on Saturday and noon on Sunday.

Age (years)	2	6	5	3	12	9	2	10	5	10	7	11	12	3
No. of hours sleep	15	13.1	13.2	14.8	10.1	11.8	15.6	11.6	13.5	11.8	12.8	10.2	9.5	14

a) On the scatter diagram, plot the information from the table. (4 marks)

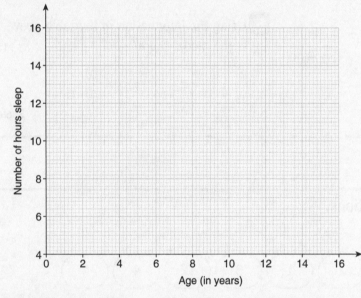

b) Describe the correlation between the age of the children in years and the total number of hours sleep they had. (2 marks)

....

....

c) Draw a line of best fit on your diagram. (1 mark)

d) Estimate the total number of hours sleep for a 4-year-old child. (2 marks)

....

e) Explain why the line of best fit only gives an estimate for the number of hours slept. (2 marks)

....

....

Score   / 11

How well did you do? ✗ 1–4 Try again   5–10 Getting there   11–15 Good work   16–21 Excellent! ✓

For more information on this topic, see pages 92–93 of your Success Guide.

77

Averages 1

A

Choose just one answer, a, b, c or d.

1 What is the mean of this set of data?
2, 7, 1, 4, 2, 6, 2, 5, 2, 6   (1 mark)

  a) 4.2   b) 3.6
  c) 3.7   d) 3.9

2 What is the median value of the set of data used in question 1?   (1 mark)

  a) 2   b) 3
  c) 4   d) 5

3 What is the range of this set of data?
2, 7, 1, 4, 11, 9, 6   (1 mark)

  a) 1   b) 6
  c) 11   d) 10

4 A die is thrown and the scores are noted. The results are shown in the table below. What is the mean die score? (C)   (1 mark)

Die score	1	2	3	4	5	6
Frequency	12	15	10	8	14	13

  a) 5   b) 3
  c) 4   d) 3.5

5 Using the information in the table above, what is the modal score?   (1 mark)

  a) 15   b) 2
  c) 4   d) 8

Score   / 5

B

Answer all parts of the questions.

1 Here are some number cards:

(8) (7) (11) (4) (2) (1) (3) (12) (4) (4)

Decide whether the following statements, which refer to the number cards above, are true or false.

a) The range of the number cards is 1–11. .... (1 mark)

b) The mean of the number cards is 5.6. .... (1 mark)

c) The median of the number cards is 5. .... (1 mark)

d) The mode of the number cards is 4. .... (1 mark)

2 A baked beans factory claims that 'On average, a tin of baked beans contains 141 beans.'

In order to check the accuracy of this claim, a sample of 20 tins was taken and the number of beans in each tin counted. (C)

Number of beans	137	138	139	140	141	142	143	144
Number of tins	1	1	1	2	5	4	4	2

a) Calculate the mean number of beans per tin. .... (2 marks)

b) Explain briefly whether you think the manufacturer is justified in making its claim.   (1 mark)

....

3 The mean of 7, 9, 10, 18, x and 17 is 13. What is the value of x? .... (C)(2 marks)

Score   / 9

(C) *Indicates that a calculator may be used*

C

These are GCSE-style questions. Answer all parts of the questions. Show your workings (on separate paper if necessary) and include the correct units in your answers.

1 Some students took a test. The table gives information about their marks in the test.

Mark	Frequency
3	2
4	5
5	11
6	2

Work out the mean mark. Ⓒ

(3 marks)

....

....

2 Simon has sat three examinations. His mean score is 65. To pass the unit, he needs to get an average of 69. What score must he get in the fourth and final examination to pass the unit? Ⓒ

(3 marks)

....

....

3 A company employs 3 women and 7 men.

The mean weekly wage of the 10 employees is £464.

The mean weekly wage of the 3 women is £520.

Calculate the mean weekly wage of the 7 men. Ⓒ

(4 marks)

....

....

4 Find the three-point moving average for the following data: Ⓒ

2   5   3   6   2   1   4

(5 marks)

....

....

....

....

Score   / 15

How well did you do?   ✗ 1–10 Try again   11–16 Getting there   17–22 Good work   23–29 Excellent! ✓

For more information on this topic, see pages 94–95 of your Success Guide.

79

Averages 2

A Choose just one answer, a, b, c or d.

The following questions are based on the information given in the table below about the time taken in seconds to swim 50 metres.

Time (seconds)	Frequency (f)
$0 \leq t < 30$	1
$30 \leq t < 60$	2
$60 \leq t < 90$	4
$90 \leq t < 120$	6
$120 \leq t < 150$	7
$150 \leq t < 180$	2

1 How many people swam 50 metres in less than 60 seconds? **(1 mark)**

a) 2
c) 3

b) 4
d) 6

2 Which of the intervals is the modal class? **(1 mark)**

a) $60 \leq t < 90$
c) $30 \leq t < 60$

b) $120 \leq t < 150$
d) $90 \leq t < 120$

3 Which of the class intervals contains the median value? **(1 mark)**

a) $90 \leq t < 120$
c) $120 \leq t < 150$

b) $150 \leq t < 180$
d) $60 \leq t < 90$

4 What is the estimate for the mean time taken to swim 50 metres? **(C)** **(1 mark)**

a) 105 seconds
c) 100 seconds

b) 385 seconds
d) 125 seconds

Score / 4

B Answer all parts of the questions.

1 The length of some seedlings are shown in the table below.

Length (mm)	Number of seedlings
$0 \leq L < 10$	3
$10 \leq L < 20$	5
$20 \leq L < 30$	9
$30 \leq L < 40$	2
$40 \leq L < 50$	1

Calculate an estimate for the mean length of the seedlings. **(4 marks)**

Mean = ... mm

2 The stem-and-leaf diagram shows the marks gained by some students in a mathematics examination.

Stem	Leaf
1	2 5 7
2	6 9
3	4 5 5 7
4	2 7 7 7 7
5	2

Stem = 10 marks
Key: 1 | 2 = 12 marks

Using the stem-and-leaf diagram, calculate:

a) the mode. .... **(1 mark)**

b) the median. .... **(1 mark)**

c) the range. .... **(1 mark)**

Score / 7

(C) Indicates that a calculator may be used

C These are GCSE-style questions. Answer all parts of the questions. Show your workings (on separate paper if necessary) and include the correct units in your answers.

1 A psychologist records the times, to the nearest minute, taken by 20 students to complete a logic problem.

Here are the results.

12	22	31	36	35	14	27	23	19	25
15	17	15	27	32	38	41	18	27	18

Draw a stem-and-leaf diagram to show this information. (4 marks)

2 John asks 100 people how much they spent last year on newspapers. The results are given in the table below.

Amount £ (x)	Frequency
$0 \leq x < 10$	12
$10 \leq x < 20$	20
$20 \leq x < 30$	15
$30 \leq x < 40$	18
$40 \leq x < 50$	14
$50 \leq x < 60$	18
$60 \leq x < 70$	3

a) Calculate an estimate of the mean amount spent on newspapers. **C** (4 marks)

....

b) Explain briefly why this value of the mean is only an estimate. (1 mark)

....

c) Calculate the class interval in which the median lies. **C** (2 marks)

....

Score / 11

How well did you do? ✗ 1–6 Try again   7–11 Getting there   12–16 Good work   17–22 Excellent! ✓

For more information on this topic, see pages 96–97 of your Success Guide.

81

Cumulative frequency graphs

A Choose just one answer, a, b, c or d.

The data below show the number of letters delivered to each of the 15 houses in Whelan Avenue (arranged in order of size).

    0, 0, 1, 1, 1, 1, 1, 1, 2, 2, 2, 3, 4, 5, 5

Use the information above to answer these questions.

1 What is the median number of letters delivered?   (1 mark)

   a) 0   b) 2   c) 1   d) 5

2 What is the lower quartile for the number of letters delivered?   (1 mark)

   a) 0   b) 2   c) 3   d) 1

3 What is the upper quartile for the number of letters delivered?   (1 mark)

   a) 0   b) 2
   c) 3   d) 1

4 What is the interquartile range for the number of letters delivered?   (1 mark)

   a) 2   b) 3
   c) 4   d) 5   Score   / 4

B Answer all parts of the questions.

1 The table shows the examination marks of some year 10 pupils in their end-of-year mathematics examination.

Examination mark	Frequency	Cumulative frequency
0–10	4	
11–20	6	
21–30	11	
31–40	24	
41–50	18	
51–60	7	
61–70	3	

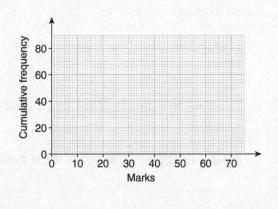

a) Complete the cumulative frequency column in the table above.   (2 marks)

b) Draw the cumulative frequency graph.   (3 marks)

c) From your graph, find the median mark. ....   (1 mark)

d) From your graph, find the interquartile range. ....   (2 marks)

e) If 16 pupils were given a grade A in the examination, what is the minimum score needed for a grade A?   (2 marks)

.... marks   Score   / 10

C These are GCSE-style questions. Answer all parts of the questions. Show your workings (on separate paper if necessary) and include the correct units in your answers.

1 The table gives information about the time, to the nearest minute, taken to run a marathon.

Time (mins)	Frequency	Cumulative frequency
$120 < t \leq 140$	1	
$140 < t \leq 160$	8	
$160 < t \leq 180$	24	
$180 < t \leq 200$	29	
$200 < t \leq 220$	10	
$220 < t \leq 240$	5	
$240 < t \leq 260$	3	

a) Complete the table to show the cumulative frequency for this data. (2 marks)

b) Draw the cumulative frequency graph for these data. (3 marks)

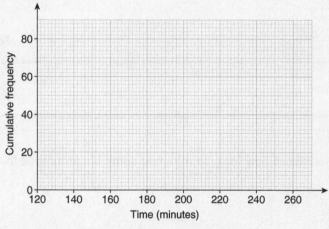

c) Use your graph to work out an estimate for:

   i) the interquartile range. .... minutes (2 marks)

   ii) the number of runners with a time of more than 205 minutes. (1 mark)

   .... runners

d) Draw a box plot for this data. (3 marks)

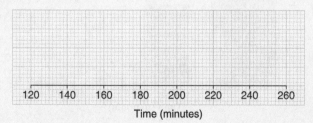

Score   / 11

How well did you do?   ✗ 1–6 Try again   7–12 Getting there   13–18 Good work   19–25 Excellent! ✓

For more information on this topic, see pages 98–99 of your Success Guide.

83

Histograms

A Choose just one answer, a, b, c or d.

The table shows the distance travelled to work by some employees. Use the information in the table to answer the questions below.

Distance (km)	Frequency
$0 \leq d < 5$	8
$5 \leq d < 15$	20
$15 \leq d < 20$	135
$20 \leq d < 30$	47
$30 \leq d < 50$	80

1 Which class interval has a frequency density of 4.7? (1 mark)

a) $0 \leq d < 5$ b) $5 \leq d < 15$
c) $15 \leq d < 20$ d) $20 \leq d < 30$

2 The frequency density for one of the class intervals is 4. Which one? (1 mark)

a) $5 \leq d < 15$ b) $30 \leq d < 50$
c) $0 \leq d < 5$ d) $15 \leq d < 20$

3 Which class interval has the highest frequency density? (1 mark)

a) $0 \leq d < 5$ b) $15 \leq d < 20$
c) $20 \leq d < 30$ d) $5 \leq d < 15$

4 Which class interval has the lowest frequency density? (1 mark)

a) $0 \leq d < 5$ b) $5 \leq d < 15$
c) $15 \leq d < 20$ d) $30 \leq d < 50$

Score / 4

B Answer all parts of the questions.

1 The table and histogram give information about how long, in minutes, some students took to complete a maths problem.

Time (in minutes)	Frequency
$0 < t \leq 5$	19
$5 < t \leq 15$	....
$15 < t \leq 20$	16
$20 < t \leq 30$	....
$30 < t \leq 45$	12

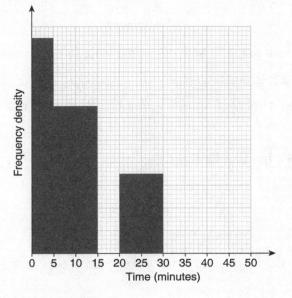

a) Use the information in the histogram to complete the table. (2 marks)
b) Use the table to complete the histogram. (2 marks)

Score / 4

C These are GCSE-style questions. Answer all parts of the questions. Show your workings (on separate paper if necessary) and include the correct units in your answers.

1 The weights of some objects are given in the table below.

Weight W (kg)	Frequency
$0 \leq W < 2$	14
$2 \leq W < 3$	8
$3 \leq W < 5$	13
$5 \leq W < 10$	14
$10 \leq W < 12$	7
$W \geq 12$	0

Draw a histogram to show the distribution of the weights of the objects. Use a scale of 2 cm to 2 kg on the weight axis. **(3 marks)**

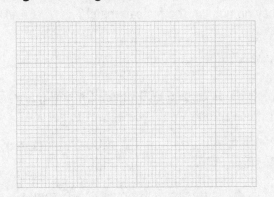

2 Pierre recorded the length, in seconds, of some advertisements shown on television in a week. His results are shown in the histogram.

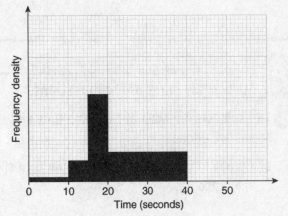

Use the information in the histogram to complete the table. **(3 marks)**

Length S (in seconds)	Frequency
$0 \leq S < 10$	....
$10 \leq S < 15$	....
$15 \leq S < 20$	21
$20 \leq S < 40$	....
$S \geq 40$	0

Score   / 6

How well did you do?   ✗ 1–2 **Try again**   3–6 **Getting there**   7–10 **Good work**   11–14 **Excellent!** ✓

For more information on this topic, see page 100–101 of your Success Guide.

85

Probability

A Choose just one answer, a, b, c or d.

1 The probability that Highbury football club win a football match is $\frac{12}{17}$. What is the probability that they do not win the football match? **(1 mark)**

a) $\frac{5}{12}$ b) $\frac{17}{29}$

c) $\frac{12}{17}$ d) $\frac{5}{17}$

2 A fair die is thrown 600 times. On how many of these throws would you expect to get a 4?

a) 40   b) 600   **(1 mark)**

c) 100   d) 580

3 A fair die is thrown 500 times. If a 6 comes up 87 times, what is the relative frequency?

a) $\frac{1}{6}$ b) $\frac{87}{500}$ **(1 mark)**

c) $\frac{10}{600}$ d) $\frac{1}{587}$

4 The probability that it snows on Christmas Day is 0.2. What is the probability that it will snow on Christmas Day in two consecutive years? **(1 mark)**

a) 0.04   b) 0.4

c) 0.2   d) 0.16

5 The probability that Fiona is in the hockey team is 0.7. The probability that she is picked for the netball team is 0.3. What is the probability that she is picked for both teams? **(1 mark)**

a) 1.0   b) 0.1

c) 0.12   d) 0.21

Score   / 5

B Answer all parts of the questions.

1 Two spinners are spun at the same time and their scores are added.

Spinner 1

3	3
2	1

Spinner 2

6	2
3	1

a) Complete the sample space diagrams to show the outcomes. **(2 marks)**

Spinner 1

Spinner 2	1	2	3	3
1	2			
2	4			
3	5			
6	8		9	

b) Find the probability of: **(3 marks)**

i) a score of 4 ....   ii) a score of 9 ....   iii) a score of 1 ....

2 The probability that Michelle finishes first in a swimming race is 0.3. Michelle swims two races. Work out the probability that Michelle wins both races. .... **(2 marks)**

3 There are 13 counters in a bag: 7 are red and the rest are white. A counter is picked at random, its colour noted and then it is replaced. A second counter is then chosen. What is the probability of choosing:

a) two red counters?   .... **(2 marks)**

b) a red and a white counter?   .... **(3 marks)**

Score   / 12

C

These are GCSE-style questions. Answer all parts of the questions. Show your workings (on separate paper if necessary) and include the correct units in your answers.

1 A bag contains different coloured beads.

The probability of taking a bead of a particular colour at random is as follows.

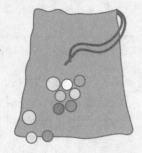

Colour	Red	White	Blue	Pink
Probability	0.25	0.1		0.3

Jackie is going to take a bead at random and then put it back in the bag.

a) i) Work out the probability that Jackie will take out a blue bead. .... (1 mark)

   ii) Write down the probability that Jackie will take out a black bead. .... (1 mark)

b) Jackie will take out a bead from the bag at random 200 times, replacing the bead each time. Work out an estimate for the number of times that Jackie takes a red bead. (2 marks)

....

2 Two fair dice are thrown together and their scores are added.

a) Work out the probability of a score of 7. .... (2 marks)

b) Work out the probability of a score of 9. .... (2 marks)

3 Kevin and Nathan challenge each other to a game of Monopoly and a game of pool. A draw is not possible in either game.

The probability that Kevin wins at Monopoly is 0.4

The probability that Nathan wins at pool is 0.7

a) Draw a probability tree diagram in the space below. (3 marks)

b) What is the probability that Nathan wins both games? .... (2 marks)

c) What is the probability that they win a game each? .... (3 marks)

Score   / 16

How well did you do?   ✗ 1–11 Try again   12–21 Getting there   22–27 Good work   28–33 Excellent! ✓

For more information on this topic, see pages 102–105 of your Success Guide.

87

Mixed GCSE-style Questions

Answer these questions. Show full working out

1 The diagram shows a circle of diameter 2.7m.
Work out the area of the circle. Give your answer
correct to 1 decimal place. (c)

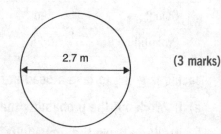

2.7 m

(3 marks)

.... m²

2 Katy sells CDs, she sells each CD for £9.20 plus
VAT@ 17.5%. She sells 127 CDs.

Work out how much money Katy receives. (c)

(4 marks)

....

3 Estimate the valve of $\frac{8.9 \times 5.2}{10.1}$ .... (2 marks)

4 Here are the ages in years of the members of a golf club.

9	42	37	28	36	44	47	43	62	19	17	36	40
56	58	32	18	41	52	42	54	38	27	29	32	51

In the space provided draw a stem and leaf diagram to show these ages. (3 marks)

5 The lines PQ and RS are parallel

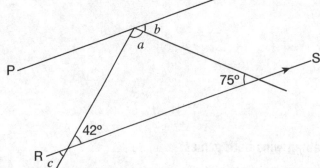

a) **Write down the value of b. Give a reason for your answer.** (2 marks)

....

b) **Write down the value of c. Give a reason for your answer.** (2 marks)

....

c) **Write down the value of a.** (2 marks)

....

(c) *Indicates that a calculator may be used*

6 Here are the first four terms of an arithmetic sequence:

5, 9, 13, 17

Find an expression, in terms of n, for the nth term of the sequence. (2 marks)

....

7 The diagram shows the position of three towns, A, B and C. Town C is due east of towns A and B. Town B is due east of A.

Town B is $3\frac{1}{3}$ miles from town A.

Town C is $1\frac{1}{4}$ miles from town B.

Calculate the number of miles between town A and town C. (3 marks)

....

8 a) **Solve $5x - 2 = 3(x + 6)$** $x =$ .... (2 marks)

b) **Solve $\dfrac{3 - 2x}{4} = 2$** $x =$ .... (2 marks)

c) i) **Factorise $x^2 - 10x + 24$**   .... (2 marks)

  ii) **Hence solve $x^2 - 10x + 24$**   .... (2 marks)

d) **Simplify $\dfrac{x^2 + 2x}{x^2 + 5x + 6}$**   .... (3 marks)

e) **Simplify these.**

  i) $p^4 \times p^6$ .... (1 mark)

  ii) $\dfrac{p^7}{p^3}$ .... (1 mark)

  iii) $\dfrac{p^4 \times p^5}{p}$ .... (1 mark)

  iv) $(p^{-\frac{1}{2}})^4$ .... (1 mark)

9 The times, in minutes, taken to finish an assault course are listed in order.

8, 12, 12, 13, 15, 17, 22, 23, 23, 27, 29

a) **Find:** (2 marks)

  i) **the lower quartile** ....

  ii) **the interquartile range** ....

b) **Draw a box plot for this data.** (3 marks)

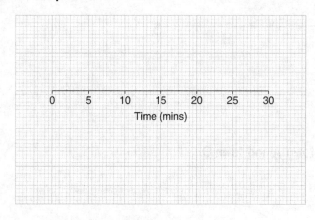

10 a) Megan bought a TV for £700. Each year the TV depreciated by 20%.

Work out the value of the TV two years after she bought it. Ⓒ (3 marks)

....

b) In a '20% off' sale, William bought a DVD player for £300.
What was the original price of the DVD player before the sale? (3 marks)

....

11 The diagram shows a right-angled triangle.

PQ = 14.2 cm.

Angle PRQ = 90°.

Angle RPQ = 38°.

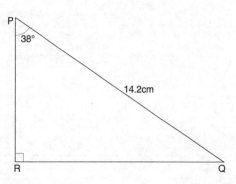

Find the length of the side QR. Give your answer to 3 significant figures. Ⓒ (3 marks)

....

Ⓒ *Indicates that a calculator may be used*

12 The table gives the times to the nearest minute to complete a puzzle.

Time (mins)	Frequency
$0 \leq t < 10$	5
$10 \leq t < 20$	12
$20 \leq t < 30$	8
$30 \leq t < 40$	5

Calculate an estimate for the mean number of minutes taken to complete the puzzle. (4 marks)

....

13 a) $b = \dfrac{a + c}{ac}$ $a = 3.2 \times 10^5$ $c = 5 \times 10^6$

Calculate the value of b. ⓒ (2 marks)

Give your answer in standard form.   ....

b) Rearrange the formula to make a the subject.   ....   (2 marks)

14

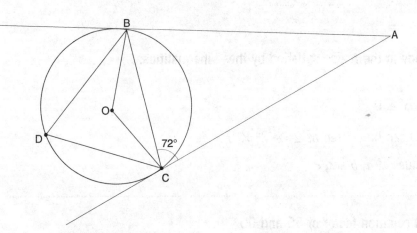

D, B and C are points on a circle with centre O.
AB and AC are tangents to the circle. Angle ACB = 72°.

a) Explain why angle OCB is 18°. (1 mark)

....

....

b) Calculate the size of angle BDC. Give reasons for your answer. (3 marks)

....

....

....

15 The diagram shows the graphs of these equations:

$x + y = 4$
$y = x - 2$

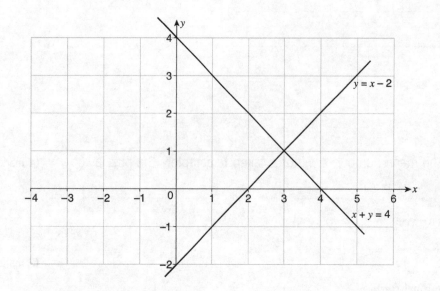

a) **Use the diagram to solve the simultaneous equations.**   (4 marks)

$x + y = 4$ $x =$ ....

$y + 2 = x$ $y =$ ....

b) **On the grid, shade in the region satisfied by these inequalities.**   (2 marks)

$x + y \leq 4$
$y \geq x - 2$ $x \geq 0$

16 a) The number 360 can be written as $2^a \times 3^b \times 5^c$.   (3 marks)

 Calculate the values of a, b and c.

 ....

b) **Find the highest common factor of 56 and 60.**   (2 marks)

 ....

c) **Find the lowest common multiple of 56 and 60.**   (2 marks)

 ....

17 a) p is an integer such that $0 < 4p \leq 13$.   (1 mark)

 List all the possible values of p.

 ....

b) **Solve the inequality $\dfrac{t + 1}{4} \leq t - 3$**   (2 marks)

 ....

18 $a = 2 + \sqrt{7}$ and $b = 2 - 3\sqrt{7}$

Simplify the following, giving your answer in the form $p + q\sqrt{7}$, where p and q are integers.

a) $a + b$ .... (3 marks)

b) a^2 .... (3 marks)

c) ab .... (3 marks)

19

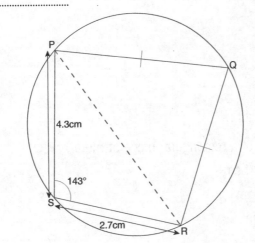

PQRS is a cyclic quadrilateral.
PS = 4.3 cm, SR = 2.7 cm, angle PSR = 143° and PQ = QR.

a) **Calculate the length of PR.** ....cm (3 marks)

b) **Calculate the length of QR.** ....cm (3 marks)

c) **Calculate the area of triangle PQR.** ....cm² (3 marks)

20 Solve the equation $\dfrac{x - 4}{x^2 - 16} + \dfrac{2}{2x - 4} = 1$ (5 marks)

Leave your answers in surd form. $x = $ ....

21 The diagram shows a child's toy, which is hollow. The height of the cone is 4 cm. The base radius of cone and hemisphere is 3 cm.

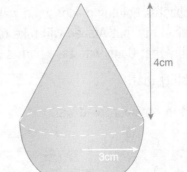

a) **Work out the total surface area of the toy.**
Give your answer as a multiple of π. (4 marks)

................................cm²

b) **The toy is made in two sizes. The large toy is three times the size of the toy above.**
What is the total surface area of the large toy? Give your answer as a multiple of π. (3 marks)

................................cm²

22

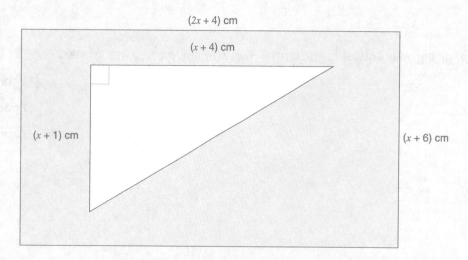

A triangular piece of metal is cut out of a rectangular piece of metal.

The length of the rectangle is $(2x + 4)$ cm

The width of the rectangle is $(x + 6)$ cm

The height of the triangle is $(x + 4)$ cm

The base of the triangle is $(x + 1)$ cm

The shaded region in the diagram shows the metal remaining.

The area of the shaded region is 38.5 cm².

a) Show that $x^2 + 9x - 11 = 0$.

(4 marks)

....

....

b) i) Solve the equation $x^2 + 9x - 11 = 0$.
   Give your answer correct to 3 significant figures.

(3 marks)

....

ii) Hence find the area of the triangle.

(1 mark)

....

23 Riddlington High School is holding a sponsored walk. The pupils at the school decide whether or not to take part. The probability that Afshan will take part is $\frac{2}{3}$. The probability that Bethany will take part is $\frac{4}{5}$ and the probability that Colin with take part is $\frac{1}{4}$.

Calculate the probability that:

a) all three take part in the sponsored walk

(2 marks)

....

b) exactly two of them take part in the sponsored walk

(3 marks)

....

Answers to mixed questions

1 Area $= \pi r^2$
$= \pi \times 1.35^2$
$= 5.7$ cm^2

2 $9.20 \times 1.175 = £10.81$
with VAT for each CD.
$127 \times £10.81$
$= £1372.87$

3 $\dfrac{8.9 \times 5.2}{10.1}$

$\approx \dfrac{9 \times 5}{10}$

$= 4.5$

4
0	9
1	7 8 9
2	7 8 9
3	2 2 6 6 7 8
4	0 1 2 2 3 4 7
5	1 2 4 6 8
6	2

1|7 means 17 years
stem = 10 years

5 a) $b = 75°$ since angle b and $75°$ are alternate angles.
b) $c = 42°$ since c and $42°$ are vertically opposite
c) $a = 63°$

6 $4n + 1$

7 $4\frac{7}{12}$ miles

8 a) $x = 10$
b) $x = -2.5$
c) i) $(x - 4)(x - 6)$
ii) $x = 4$ and $x = 6$
d) $\dfrac{x(x + 2)}{(x + 2)(x + 3)} = \dfrac{x}{x + 3}$
e) i) p^{10} ii) p^4 iii) p^8 iv) $p^{-2} = \frac{1}{p^2}$

9 a) i) lower quartile = 12
ii) interquartile range = 11

9 b)

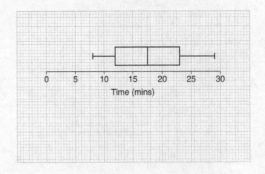

10 a) £448   b) £375

11 8.74 cm

12 19.3̇ minutes

13 a) 3.325×10^{-6} b) $a = \dfrac{c}{bc - 1}$

14 a) The radius and tangent meet at 90°.
Since angle ACB = 72° then angle OCB = 90° − 72° = 18°.
b) Angle OCB = angle OBC = 18°. Angle BOC = 180° − (2 × 18°) = 144°.
Angle BDC = 144° ÷ 2 = 72°, since the angle subtended at the centre is twice the angle at the circumference.

15 a) $x = 3, y = 1$
b)

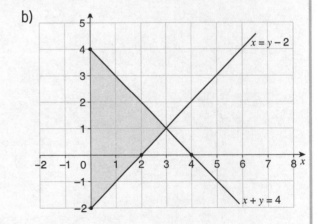

16 a) $a = 3, b = 2, c = 1$ b) 4   c) 840

17 a) 1, 2, 3   b) $t \geq \dfrac{13}{3}$

18 a) $4 - 2\sqrt{7}$ b) $11 + 4\sqrt{7}$
c) $-17 - 4\sqrt{7}$

19 a) 6.66 cm (3 s.f.)

b) 10.49 cm (3 s.f.)

c) 33.1 cm^2 (3 s.f.)

20 $\dfrac{x-4}{x^2-16} + \dfrac{2}{2x-4} = 1$

$\dfrac{x-4}{(x-4)(x+4)} + \dfrac{2}{2(x-2)} = 1$

$\dfrac{1}{(x+4)} + \dfrac{1}{(x-2)} = 1$

$x - 2 + x + 4 = 1\,(x+4)(x-2)$

$2x + 2 = x^2 + 2x - 8$

$x^2 + 2x - 8 - 2x - 2 = 0$

$x^2 - 10 = 0$

$x = \pm\sqrt{10}$

$x = +\sqrt{10}$ or $x = -\sqrt{10}$

21 a) Curved surface area of cone $= \pi rl$

$\pi \times 3 \times 5 = 15\pi$

Curved surface area of hemisphere $= \dfrac{4\pi r^2}{2}$

$2\pi \times 9 = 18\pi$

Total curved surface area $= 33\pi$

b) Linear scale factor $= 3$

Area scale factor $= 3^2 = 9$

Curved surface area of larger toy $= 297\pi$

22 a) $(2x+4)(x+6) - \frac{1}{2} \times (x+1)(x+4) = 38.5$

$2x^2 + 16x + 24 - \dfrac{(x^2+5x+4)}{2} = 38.5$

$4x^2 + 32x + 48 - x^2 - 5x - 4 = 77$

$3x^2 + 27x + 44 = 77$

$3x^2 + 27x - 33 = 0$

$(\div 3) \therefore x^2 + 9x - 11 = 0$

b) i) $x = 1.09$ or $x = -10.09$, but $x > 0$

ii) Area of triangle $= \frac{1}{2} \times 2.09 \times 5.09$

$= 5.319\,05$

$= 5.32$ cm^2 (3 s.f.)

23 a) $\dfrac{2}{15}$ 　　　　b) $\dfrac{1}{2}$